Stellarium User Guide

A catalogue record for this book is available from the Hong Kong Public Libraries.

Published in Hong Kong by Samurai Media Limited.

Email: info@samuraimedia.org

ISBN 978-988-8406-79-1

Minor modifications for publication Copyright 2016 Samurai Media Limited.

Background Cover Image by https://www.flickr.com/people/webtreatsetc/

Contents

I   Basic Use

1   Introduction 17

1.1   **Historical notes**   17

2   Getting Started 21

2.1   **System Requirements**   21

2.1.1   Minimum 21

2.1.2   Recommended 21

2.2   **Downloading**   22

2.3   **Installation**   22

2.3.1   Windows 22

2.3.2   OS X 22

2.3.3   Linux 22

2.4   **Running Stellarium**   22

2.4.1   Windows 22

2.4.2   OS X 23

2.4.3   Linux 23

2.5   **Troubleshooting**   23

3   A First Tour 25

3.1   **Time Travel**   26

3.2   **Moving Around the Sky**   27

3.3	**The Main Tool Bar**	**28**
3.4	**Taking Screenshots**	**30**

4 The User Interface 31

4.1	**Setting the Date and Time**	**31**
4.2	**Setting Your Location**	**32**
4.3	**The Configuration Window**	**33**
4.3.1	The Main Tab	33
4.3.2	The Information Tab	33
4.3.3	The Navigation Tab	33
4.3.4	The Tools Tab	33
4.3.5	The Scripts Tab	37
4.3.6	The Plugins Tab	37
4.4	**The View Settings Window**	**37**
4.4.1	Sky Tab	37
4.4.2	DSO Tab	39
4.4.3	Markings Tab	39
4.4.4	Landscape Tab	40
4.4.5	Starlore Tab	41
4.5	**The Object Search Window**	**42**
4.6	**The AstroCalc Window**	**44**
4.7	**Help Window**	**44**
4.7.1	Editing Keyboard Shortcuts	46

II Advanced Use

5 Files and Directories 49

5.1	**Directories**	**49**
5.1.1	Windows	49
5.1.2	Mac OS X	50
5.1.3	Linux	50
5.2	**Directory Structure**	**50**
5.3	**The Main Configuration File**	**51**
5.4	**Getting Extra Data**	**51**
5.4.1	Alternative Planet Ephemerides: DE430, DE431	51

6 Command Line Options 53

6.1	**Examples**	**55**

7 Landscapes 57

7.1	**Stellarium Landscapes**	**57**
7.1.1	Location information	58

7.1.2	Polygonal landscape	59
7.1.3	Spherical landscape	60
7.1.4	High resolution ("Old Style") landscape	61
7.1.5	Fisheye landscape	65
7.1.6	Description	66
7.1.7	Gazetteer	67
7.2	**Creating Panorama Photographs for Stellarium**	**68**
7.2.1	Panorama Photography	68
7.2.2	Hugin Panorama Software	69
7.2.3	Regular creation of panoramas	70
7.3	**Panorama Postprocessing**	**73**
7.3.1	The GIMP	73
7.3.2	ImageMagick	74
7.3.3	Final Calibration	76
7.3.4	Artificial Panoramas	78
7.3.5	Nightscape Layer	78
7.4	**Other recommended software**	**79**
7.4.1	IrfanView	79
7.4.2	FSPViewer	79
7.4.3	Clink	79
7.4.4	Cygwin	80
7.4.5	GNUWin32	80
8	**Deep-Sky Objects**	**81**
8.1	**Stellarium DSO Catalog**	**81**
8.1.1	Modifying catalog.dat	82
8.1.2	Modifying names.dat	84
8.1.3	Modifying textures.json	84
8.2	**Adding Extra Nebulae Images**	**85**
8.2.1	Preparing a photo for inclusion to the `textures.json` file	85
8.2.2	Plate Solving	87
8.2.3	Processing into a `textures.json` insert	87
9	**Adding Sky Cultures**	**89**
9.1	**Basic Information**	**89**
9.2	**Skyculture Description Files**	**90**
9.3	**Constellation Names**	**90**
9.4	**Star Names**	**90**
9.5	**Planet Names**	**91**
9.6	**Stick Figures**	**91**
9.7	**Constellation Borders**	**91**
9.8	**Constellation Artwork**	**91**
9.9	**Seasonal Rules**	**92**

9.10   **Publish Your Work**   **92**

III   Extending Stellarium

10   Plugins 95
10.1   **Enabling plugins**   **95**
10.2   **Data for plugins**   **95**

11   Interface Extensions 97
11.1   **Angle Measure Plugin**   **97**
11.2   **Compass Marks Plugin**   **98**
11.3   **Equation of Time Plugin**   **99**
11.3.1 Section $\big[$EquationOfTime$\big]$ in config.ini file 99
11.4   **Field of View Plugin**   **100**
11.4.1 Section $\big[$FOV$\big]$ in config.ini file 100
11.5   **Pointer Coordinates Plugin**   **101**
11.5.1 Section $\big[$PointerCoordinates$\big]$ in config.ini file 101
11.6   **Text User Interface**   **102**
11.6.1 Using the Text User Interface 102
11.6.2 TUI Commands 102
11.6.3 Section $\big[$tui$\big]$ in config.ini file 105
11.7   **Remote Control**   **106**
11.7.1 Using the plugin 106
11.7.2 Remote Control Web Interface 107
11.7.3 Remote Control Commandline API 107
11.7.4 Developer information 107
11.8   **Solar System Editor Plugin**   **108**
11.9   **Timezone Configuration Plugin**   **109**

12   Object Catalog Plugins 111
12.1   **Bright Novae Plugin**   **111**
12.1.1 Section $\big[$Novae$\big]$ in config.ini file 111
12.1.2 Format of bright novae catalog 112
12.1.3 Light curves 113
12.2   **Historical Supernovae Plugin**   **114**
12.2.1 List of supernovae in default catalog 114
12.2.2 Light curves 115
12.2.3 Section $\big[$Supernovae$\big]$ in config.ini file 116
12.2.4 Format of historical supernovae catalog 117

12.3	**Exoplanets Plugin**	**118**
12.3.1	Potential habitable exoplanets	118
12.3.2	Proper names	119
12.3.3	Section [Exoplanets] in config.ini file	121
12.3.4	Format of exoplanets catalog	122
12.4	**Pulsars Plugin**	**124**
12.4.1	Section [Pulsars] in config.ini file	124
12.4.2	Format of pulsars catalog	125
12.5	**Quasars Plugin**	**126**
12.5.1	Section [Quasars] in config.ini file	126
12.5.2	Format of quasars catalog	127
12.6	**Meteor Showers Plugin**	**128**
12.6.1	Terms	128
12.6.2	Section [MeteorShowers] in config.ini file	129
12.6.3	Format of Meteor Showers catalog	130
12.6.4	Further Information	131
12.7	**Navigational Stars Plugin**	**132**
12.7.1	Section [NavigationalStars] in config.ini file	132
12.8	**Satellites Plugin**	**133**
12.8.1	Satellite Properties	133
12.8.2	Satellite Catalog	133
12.8.3	Configuration	134
12.8.4	Sources for TLE data	134
12.9	**ArchaeoLines Plugin**	**135**
12.9.1	Introduction	135
12.9.2	Characteristic Declinations	135
12.9.3	Azimuth Lines	137
12.9.4	Configuration Options	137
13	**Scenery3d – 3D Landscapes**	**139**
13.1	**Introduction**	**139**
13.2	**Usage**	**139**
13.3	**Hardware Requirements & Performance**	**140**
13.3.1	Performance notes	141
13.4	**Model Configuration**	**141**
13.4.1	Exporting OBJ from Sketchup	141
13.4.2	Notes on OBJ file format limitations	142
13.4.3	Configuring OBJ for Scenery3d	143
13.4.4	Concatenating OBJ files	147
13.4.5	Working with non-georeferenced OBJ files	147
13.4.6	Rotating OBJs with recognized survey points	148
13.5	**Predefined views**	**148**

14 Stellarium at the Telescope 151

14.1	**Oculars Plugin**	**151**
14.2	**TelescopeControl Plugin**	**152**
14.2.1	Abilities and limitations	152
14.2.2	Using this plug-in	152
14.2.3	Main window ('Telescopes')	152
14.2.4	Telescope configuration window	153
14.2.5	Supported devices	155
14.3	**StellariumScope plugin**	**156**
14.4	**Other telescope servers and Stellarium**	**156**
14.5	**Observability Plugin**	**158**

15 Scripting 161

15.1	**Introduction**	**161**
15.2	**Script Console**	**162**
15.3	**Includes**	**162**
15.4	**Minimal Scripts**	**162**
15.5	**Example: Retrograde motion of Mars**	**162**
15.5.1	Script header...	162
15.5.2	A body of script...	163
15.6	**More Examples**	**165**

IV Practical Astronomy

16 Astronomical Concepts 169

16.1	**The Celestial Sphere**	**169**
16.2	**Coordinate Systems**	**170**
16.2.1	Altitude/Azimuth Coordinates	170
16.2.2	Right Ascension/Declination Coordinates	171
16.2.3	Ecliptical Coordinates	173
16.2.4	Galactic Coordinates	174
16.3	**Units**	**174**
16.3.1	Distance	174
16.3.2	Time	174
16.3.3	Julian Day Number	175
16.3.4	Angles	179
16.3.5	The Magnitude Scale	180
16.3.6	Luminosity	180
16.4	**Precession**	**181**
16.5	**Parallax**	**181**
16.5.1	Geocentric and Topocentric Observations	183
16.5.2	Stellar Parallax	183

16.6   **Proper Motion**   **183**

17   Astronomical Phenomena 185

17.1   **The Sun**   **185**

17.2   **Stars**   **185**

17.2.1 Multiple Star Systems 186

17.2.2 Constellations 186

17.2.3 Star Names 187

17.2.4 Spectral Type & Luminosity Class 188

17.2.5 Variable Stars 190

17.3   **Our Moon**   **190**

17.3.1 Phases of the Moon 190

17.4   **The Major Planets**   **192**

17.4.1 Terrestrial Planets 192

17.4.2 Jovian Planets 192

17.5   **The Minor Bodies**   **193**

17.5.1 Asteroids 193

17.5.2 Comets 193

17.6   **Meteoroids**   **193**

17.7   **Zodiacal Light and *Gegenschein***   **194**

17.8   **The Milky Way**   **194**

17.9   **Nebulae**   **194**

17.9.1 The Messier Objects 195

17.10   **Galaxies**   **195**

17.11   **Eclipses**   **195**

17.11.1 Solar Eclipses 195

17.11.2 Lunar Eclipses 196

17.12   **Observing Hints**   **196**

17.13   **Atmospheric effects**   **197**

17.13.1 Atmospheric Refraction 197

17.13.2 Atmospheric Extinction 197

17.13.3 Light Pollution 197

18   A Little Sky Guide 201

18.1   **Dubhe and Merak, The Pointers**   **201**

18.2   **M31, Messier 31, The Andromeda Galaxy**   **201**

18.3   **The Garnet Star, μ Cephei**   **202**

18.4   **4 and 5 Lyrae, ε Lyrae**   **202**

18.5   **M13, Hercules Cluster**   **202**

18.6   **M45, The Pleiades, The Seven Sisters**   **202**

18.7   **Algol, The Demon Star, β Persei**   **203**

18.8   **Sirius, α Canis Majoris**   **203**

18.9	M44, The Beehive, Praesepe	203
18.10	27 Cephei, δ Cephei	203
18.11	M42, The Great Orion Nebula	203
18.12	La Superba, Y Canum Venaticorum, HIP 62223	204
18.13	52 and 53 Bootis, v^1 and v^2 Bootis	204
18.14	PZ Cas, HIP 117078	204
18.15	VV Cephei, HIP 108317	204
18.16	AH Scorpii, HIP 84071	204
18.17	Albireo, β Cygni	205
18.18	31 and 32 Cygni, o^1 and o^2 Cygni	205
18.19	The Coathanger, Brocchi's Cluster, Cr 399	205
18.20	Kemble's Cascade	206
18.21	The Double Cluster, χ and h Persei, NGC 884 and NGC 869	206
18.22	Large Magellanic Cloud, PGC 17223	206
18.23	Tarantula Nebula, C 103, NGC 2070	206
18.24	Small Magellanic Cloud, NGC 292, PGC 3085	207
18.25	ω Centauri cluster, C 80, NGC 5139	207
18.26	47 Tucanae, C 106, NGC 104	208
18.27	The Coalsack Nebula, C 99	208
18.28	Mira, o Ceti, 68 Cet	208
18.29	α Persei Cluster, Cr 39, Mel 20	208
18.30	M7, The Ptolemy Cluster	209
18.31	M24, The Sagittarius Star Cloud	209
18.32	IC 4665, The Summer Beehive Cluster	210
18.33	The E Nebula, Barnard 142 and 143	210
19	Exercises	211
19.1	Find M31 in Binoculars	211
19.1.1	Simulation	211
19.1.2	For Real	211
19.2	Handy Angles	211
19.3	Find a Lunar Eclipse	212
19.4	Find a Solar Eclipse	212
19.5	Find a retrograde motion of Mars	212
19.6	Analemma	212
19.7	Transit of Venus	212
19.8	Transit of Mercury	213
19.9	Triple shadows on Jupiter	213
19.10	Jupiter without satellites	213
19.11	Mutual occultations of planets	213
19.12	The proper motion of stars	213

V   Appendices

A   Default Hotkeys 217

A.1	**Display Options**	**217**
A.2	**Miscellaneous**	**218**
A.3	**Movement and Selection**	**218**
A.4	**Date and Time**	**218**
A.5	**Scripts**	**219**
A.6	**Windows**	**219**
A.7	**Plugins**	**219**
A.7.1	Angle Measure	219
A.7.2	ArchaeoLines	220
A.7.3	Compass Marks	220
A.7.4	Equation of Time	220
A.7.5	Exoplanets	220
A.7.6	Field of View	220
A.7.7	Meteor Showers	220
A.7.8	Oculars	221
A.7.9	Pulsars	221
A.7.10	Quasars	221
A.7.11	Satellites	221
A.7.12	Scenery3d: 3D landscapes	221
A.7.13	Solar System Editor	221
A.7.14	Telescope Control	222

B   The Bortle Scale of Light Pollution 223

B.1	**Excellent dark sky site**	**223**
B.2	**Typical truly dark site**	**223**
B.3	**Rural sky**	**223**
B.4	**Rural/suburban transition**	**224**
B.5	**Suburban sky**	**224**
B.6	**Bright suburban sky**	**224**
B.7	**Suburban/urban transition**	**224**
B.8	**City sky**	**224**
B.9	**Inner City sky**	**225**

C   Star Catalogues 227

C.1	**Stellarium's Sky Model**	**227**
C.1.1	Zones	227
C.2	**Star Catalogue File Format**	**227**
C.2.1	General Description	227
C.2.2	File Sections	228
C.2.3	Record Types	229

C.3	**Variable Stars**	**232**
C.3.1	Variable Star Catalog File Format	232
C.3.2	GCVS Variability Types	232
C.4	**Double Stars**	**247**
C.4.1	Double Star Catalog File Format	247
C.5	**Cross-Identification Data**	**248**
C.5.1	Cross-Identification Catalog File Format	248
D	**Configuration Files**	**249**
D.1	**Program Configuration**	**249**
D.1.1	[astro]	249
D.1.2	[color]	252
D.1.3	[custom_selected_info]	254
D.1.4	[custom_time_correction]	254
D.1.5	[devel]	255
D.1.6	[dso_catalog_filters]	255
D.1.7	[dso_type_filters]	255
D.1.8	[gui]	256
D.1.9	[init_location]	257
D.1.10	[landscape]	258
D.1.11	[localization]	258
D.1.12	[main]	259
D.1.13	[navigation]	259
D.1.14	[plugins_load_at_startup]	260
D.1.15	[projection]	261
D.1.16	[proxy]	262
D.1.17	[scripts]	262
D.1.18	[search]	262
D.1.19	[spheric_mirror]	262
D.1.20	[stars]	263
D.1.21	[tui]	264
D.1.22	[video]	264
D.1.23	[viewing]	264
D.2	**Solar System Configuration File**	**267**
D.2.1	Planet section	267
D.2.2	Moon section	268
D.2.3	Minor Planet section	270
D.2.4	Comet section	270

D.2.5   Solar System Observer 272

E   Accuracy 275

E.1   **Planetary Positions**   **275**

E.2   **Minor Bodies**   **276**

E.3   **Precession and Nutation**   **276**

E.4   **Planet Axes**   **276**

F   Contributors 277

F.1   **How you can help**   **277**

G   GNU Free Documentation License 279

G.1   **PREAMBLE**   **279**

G.2   **APPLICABILITY AND DEFINITIONS**   **279**

G.3   **VERBATIM COPYING**   **281**

G.4   **COPYING IN QUANTITY**   **281**

G.5   **MODIFICATIONS**   **281**

G.6   **COMBINING DOCUMENTS**   **283**

G.7   **COLLECTIONS OF DOCUMENTS**   **283**

G.8   **AGGREGATION WITH INDEPENDENT WORKS**   **283**

G.9   **TRANSLATION**   **283**

G.10   **TERMINATION**   **284**

G.11   **FUTURE REVISIONS OF THIS LICENSE**   **284**

Bibliography 285

Index 291

1   Introduction 17

1.1   Historical notes

2   Getting Started 21

2.1   System Requirements
2.2   Downloading
2.3   Installation
2.4   Running Stellarium
2.5   Troubleshooting

3   A First Tour 25

3.1   Time Travel
3.2   Moving Around the Sky
3.3   The Main Tool Bar
3.4   Taking Screenshots

4   The User Interface 31

4.1   Setting the Date and Time
4.2   Setting Your Location
4.3   The Configuration Window
4.4   The View Settings Window
4.5   The Object Search Window
4.6   The AstroCalc Window
4.7   Help Window

1. Introduction

Stellarium is a software project that allows people to use their home computer as a virtual planetarium. It calculates the positions of the Sun and Moon, planets and stars, and draws how the sky would look to an observer depending on their location and the time. It can also draw the constellations and simulate astronomical phenomena such as meteor showers or comets, and solar or lunar eclipses.

Stellarium may be used as an educational tool for teaching about the night sky, as an observational aid for amateur astronomers wishing to plan a night's observing or even drive their telescopes to observing targets, or simply as a curiosity (it's fun!). Because of the high quality of the graphics that Stellarium produces, it is used in some real planetarium projector products and museum projection setups. Some amateur astronomy groups use it to create sky maps for describing regions of the sky in articles for newsletters and magazines, and the exchangeable sky cultures feature invites its use in the field of Cultural Astronomy research and outreach.

Stellarium is still under development, and by the time you read this guide, a newer version may have been released with even more features than those documented here. Check for updates to Stellarium at the Stellarium website[1].

If you have questions and/or comments about this guide, or about Stellarium itself, visit the Stellarium site at LaunchPad[2] or our SourceForge forums[3].

1.1 Historical notes

Fabien Chéreau started the project during the summer 2000, and throughout the years found continuous support by a small team of enthusiastic developers.

Here is a list of past and present major contributors sorted roughly by date of arrival on the project:

---

[1]http://stellarium.org
[2]https://launchpad.net/stellarium
[3]https://sourceforge.net/p/stellarium/discussion/278769/

Fabien Chéreau original creator, maintainer, general development

Matthew Gates maintainer, original user guide, user support, general development

Johannes Gajdosik astronomical computations, large star catalogs support

Johan Meuris GUI design, website creation, drawings of our 88 Western constellations

Nigel Kerr Mac OSX port

Rob Spearman funding for planetarium support

Barry Gerdes user support, tester, Windows support. Barry passed away in October 2014 at age 80. He was a major contributor on the forums, wiki pages and mailing list where his good will and enthusiasm is strongly missed. Version 0.15 of Stellarium is dedicated in his memory. RIP Barry.

Timothy Reaves ocular plugin

Bogdan Marinov GUI, telescope control, other plugins

Diego Marcos SVMT plugin

Guillaume Chéreau display, optimization, Qt upgrades

Alexander Wolf maintainer, DSO catalogs, user guide, general development

Georg Zotti astronomical computations, Scenery 3D and ArchaeoLines plugins, general development, user guide, user support

Marcos Cardinot MeteorShowers plugin

Florian Schaukowitsch Scenery 3D plugin, Remote Control plugin, Qt/OpenGL internals

Unfortunately time is evolving, and most members of the original development team are no longer able to devote most of their spare time to the project (some are still available for limited work which requires specific knowledge about the project).

As of 2016, the project's maintainer is Alexander Wolf, doing most maintenance and regular releases. Major new features are contributed mostly by Georg Zotti and his team focussing on extensions of Stellarium's applicability in the fields of historical and cultural astronomy (which means Stellarium is getting more accurate), but also on graphic items like comet tails or the Zodiacal Light.

A detailed track of development can be found in the `ChangeLog` file in the installation folder. A few important milestones for the project:

2000 first lines of code for the project

2001-06 first public mention (and users feedbacks!) of the software on the French newsgroup `fr.sci.astronomie.amateur` [4]

2003-01 Stellarium reviewed by Astronomy magazine

2003-07 funding for developing planetarium features (fisheye projection and other features)

2005-12 use accurate (and fast) planetary model

2006-05 Stellarium "Project Of the Month" on SourceForge

2006-08 large stars catalogs

2007-01 funding by ESO for development of professional astronomy extensions (VirGO)

2007-04 developers' meeting near Munich, Germany

2007-05 switch to the Qt library as main GUI and general purpose library

2009-09 plugin system, enabling a lot of new development

2010-07 Stellarium ported on Maemo mobile device

2010-11 artificial satellites plugin

2014-06 high quality satellites and Saturn rings shadows, normal mapping for moon craters

2014-07 V0.13: adapt to OpenGL evolutions in the Qt framework, now requires more modern graphic hardware than earlier versions

---

[4]`https://groups.google.com/d/topic/fr.sci.astronomie.amateur/OT7K8yogRlI/discussion`

2015-04 V0.13.3: Scenery 3D plugin

2015-10 V0.14.0: Accurate precession

2016-07 V0.15: Remote Control plugin

Stellarium has been kindly supported by ESA in their Summer of Code in Space initiatives, which resulted in better planetary rendering (2012), the Meteor Showers plugin (2013) and the web-based remote control and an alternative solution for planetary positions based on the DE430/DE431 ephemeris (2015).

This guide is based on the user guide written by Matthew Gates for version 0.10 around 2008. The guide was then ported to the Stellarium wiki and continuously updated by Barry Gerdes and Alexander Wolf up to version 0.12.

The user documentation has been developed further on the Stellarium wiki for some time, but without Barry started to fall out of sync with the actual program. We (Alexander and Georg) have ported the texts back to LaTeX and updated and added information where necessary. We feel now that a single book may be the better format for offline reading. The PDF version of this guide has a clickable table of contents and clickable hyperlinks.

This new edition of the guide will not contain notes about using earlier versions than 0.13 or using very outdated hardware. Some references to previous version may still be made for completeness, but if you are using earlier versions for particular reasons, please use the older guides.

2. Getting Started

System Requirements

Stellarium has been seen to run on most systems where Qt5 is available, from tiny ARM computers like the Raspberry Pi 2[1] or Odroid C1 to big museum installations with multiple projectors. The most important hardware requirement is a contemporary graphics subsystem.

Minimum

- Linux/Unix; Windows 7 and later (It may run on Vista, but unsupported. A special version for XP is still available); OS X 10.8.5 and later
- 3D graphics card which supports OpenGL 3.0 and GLSL 1.3 (2008 GeForce 8xxx and later, ATI/AMD Radeon HD-2xxx and later; Intel HD graphics (Core-i 2xxx and later)) or OpenGL ES 2.0 and GLSL ES 1.0 (e.g., ARM SBCs like Raspberry Pi 2). On Windows, some older cards may be supported via ANGLE when they support DirectX10.
- 512 MB RAM
- 250 MB free on disk

Recommended

- Linux/Unix; Windows 7 and later; OS X 10.8.5 and later
- 3D graphics card which supports OpenGL 3.3 and above and GLSL1.3 and later
- 1 GB RAM or more
- 1.5 GB free on disk (About 3GB extra required for the optional DE430/DE431 files).

A dark room for realistic rendering — details like the Milky Way, Zodiacal Light or star twinkling can't be seen in a bright room.

---

[1]As of spring 2016, you need to enable the experimental OpenGL driver and compile Stellarium from sources.

2.2 Downloading

Download the correct package for your operating system directly from the main page,
`http://stellarium.org`.

2.3 Installation

2.3.1 Windows

1. Double click on the installer file you downloaded:
   - `stellarium-0.15.0-win64.exe` for 64-bit Windows 7 and later.
   - `stellarium-0.15.0-win32.exe` for 32-bit Windows 7 and later.
   - `stellarium-0.15.0-classic-win32.exe` for Windows XP and later.
2. Follow the on-screen instructions.

2.3.2 OS X

1. Locate the `Stellarium-0.15.0.dmg` file in Finder and double click on it or open it using the Disk Utility application. Now, a new disk appears on your desktop and Stellarium is in it.
2. Open the new disk and please take a moment to read the `ReadMe` file. Then drag `Stellarium` to the Applications folder.
3. Note: You should copy Stellarium to the Applications folder before running it — some users have reported problems running it directly from the disk image (`.dmg`).

2.3.3 Linux

Check if your distribution has a package for Stellarium already — if so you're probably best off using it. If not, you can download and build the source.

For Ubuntu we provide a package repository with the latest stable releases. Open a terminal and type:

```
sudo add-apt-repository ppa:stellarium/stellarium-releases
sudo apt-get update
sudo apt-get install stellarium
```

2.4 Running Stellarium

2.4.1 Windows

The Stellarium installer creates a whole list of items in the **Start Menu** under the **Programs/Stellarium** section. The list evolves over time, not all entries listed here may be installed on your system. Select one of these to run Stellarium:

Stellarium OpenGL version. This is the most efficient for modern PCs and should be used when you have installed appropriate OpenGL drivers. Note that some graphics cards are "blacklisted" by Qt to immediately run via ANGLE (Direct3D), you cannot force OpenGL in this case. This should not bother you.

Stellarium (ANGLE mode) Uses Direct3D translation of the OpenGL rendering via ANGLE library. Forces Direct3D version 9.

Stellarium (MESA mode) Uses software rendering via MESA library. This should work on any PC without dedicated graphics card.

On startup, a diagnostic check is performed to test whether the graphics hardware is capable of running. If all is fine, you will see nothing of it. Else you may see an error panel informing you that your computer is not capable of running Stellarium ("No OpenGL 2 found"), or a warning that there is only OpenGL 2.1 support. The latter means you will be able to see some graphics, but depending on the type of issue you will have some bad graphics. For example, on an Intel GMA4500 there is only a minor issue in Night Mode, while on other systems we had reports of missing planets or even crashes as soon as a planet comes into view. If you see this, try running in Direct3D 9 or MESA mode, or upgrade your system. The warning, once ignored, will not show again.

When you have found a mode that works on your system, you can delete the other links.

2.4.2 OS X

Double click on the *Stellarium* application. Add it to your **Dock** for quick access.

2.4.3 Linux

If your distribution had a package you'll probably already have an item in the GNOME or KDE application menus. If not, just open a terminal and type stellarium.

2.5 Troubleshooting

Stellarium writes startup and other diagnostic messages into a logfile. Please see section 5 where this file is located on your system. This file is *essential* in case when you feel you need to report a problem with your system which has not been found before.

If you don't succeed in running Stellarium, please see the online forum[2]. It includes FAQ (Frequently Asked Questions, also Frequently Answered Questions) and a general question section which may include further hints. Please make sure you have read and understood the FAQ before asking the same questions again.

---

[2]https://launchpad.net/stellarium

3. A First Tour

Figure 3.1: Stellarium main view. (Combination of day and night views.)

When Stellarium first starts, we see a green meadow under a sky. Depending on the time of day, it is either a day or night scene. If you are connected to the Internet, an automatic lookup will attempt to detect your approximate position.[1]

---

[1] See section 4.2 if you want to switch this off.

At the bottom left of the screen, you can see the status bar. This shows the current observer location, field of view (FOV), graphics performance in frames per second (FPS) and the current simulation date and time. If you move the mouse over the status bar, it will move up to reveal a tool bar which gives quick control over the program.

The rest of the view is devoted to rendering a realistic scene including a panoramic landscape and the sky. If the simulation time and observer location are such that it is night time, you will see stars, planets and the moon in the sky, all in the correct positions.

You can drag with the mouse on the sky to look around or use the cursor keys. You can zoom with the mouse wheel or the [Page↑] or [Page↓] keys.

Much of Stellarium can be controlled very intuitively with the mouse. Many settings can additionally be switched with shortcut keys (hotkeys). Advanced users will learn to use these shortcut keys. Sometimes a key combination will be used. For example, you can quit Stellarium by pressing [Ctrl]+[Q] on Windows and Linux, and [⌘]+[Q] on Mac OS X. For simplicity, we will show only the Windows/Linux version. We will present the default hotkeys in this guide. However, almost all hotkeys can be reconfigured to match your taste. Note that some listed shortkeys are only available as key combinations on international keyboard layouts, e.g., keys which require pressing [AltGr] on a German keyboard. These must be reconfigured, please see 4.7.1 for details.

The way Stellarium is shown on the screen is primarily governed by the menus. These are accessed by dragging the mouse to the left or bottom edge of the screen, where the menus will slide out. In case you want to see the menu bars permanently, you can press the small buttons right in the lower left corner to keep them visible.

3.1   Time Travel

When Stellarium starts up, it sets its clock to the same time and date as the system clock. However, Stellarium's clock is not fixed to the same time and date as the system clock, or indeed to the same speed. We may tell Stellarium to change how fast time should pass, and even make time go backwards! So the first thing we shall do is to travel into the future! Let's take a look at the time control buttons on the right hand ride of the tool-bar. If you hover the mouse cursor over the buttons, a short description of the button's purpose and keyboard shortcut will appear.

Button	Shortcut key	Description
◀◀	J	Decrease the rate at which time passes
▶	K	Make time pass as normal
▶▶	L	Increase the rate at which time passes
▼	8	Return to the current time & date

Table 3.1: Time Travel

OK, so lets go see the future! Click the mouse once on the increase time speed button ▶▶. Not a whole lot seems to happen. However, take a look at the clock in the status bar. You should see the time going by faster than a normal clock! Click the button a second time. Now the time is going by faster than before. If it's night time, you might also notice that the stars have started to move slightly across the sky. If it's daytime you might be able to see the sun moving (but it's less apparent than the movement of the stars). Increase the rate at which time passes again by clicking on the button a third time. Now time is really flying!

Let time move on at this fast speed for a little while. Notice how the stars move across the sky. If you wait a little while, you'll see the Sun rising and setting. It's a bit like a time-lapse movie.

Stellarium not only allows for moving forward through time – you can go backwards too! Click on the real time speed button ▶ . The stars and/or the Sun should stop scooting across the sky.

Now press the decrease time speed button ◀◀ once. Look at the clock. Time has stopped. Click the decrease time speed button four or five more times. Now we're falling back through time at quite a rate (about one day every ten seconds!).

Time Dragging

Another way to quickly change time is *time dragging*. Press Ctrl + ⬚ and slide the mouse right to go forward, or left to go backward.

Enough time travel for now. Wait until it's night time, and then click the Real time speed button. With a little luck you will now be looking at the night sky.

3.2   Moving Around the Sky

Key	Description
Cursor keys ← → ↑ ↓	Pan the view left, right, up and down
Page↑ / Page↓	Zoom in and out
Backslash (\ )	Auto-zoom out to original field of view
Left mouse button	Select an object in the sky
Right mouse button	Clear selected object
Mouse wheel	Zoom in and out
⬚	Centre view on selected object
Forward-slash (/)	Auto-zoom in to selected object

Table 3.2: Moving Around the Sky

As well as travelling through time, Stellarium lets to look around the sky freely, and zoom in and out. There are several ways to accomplish this listed in table 3.2.

Let's try it. Use the cursors to move around left, right, up and down. Zoom in a little using the Page↑ key, and back out again using the Page↓. Press the \ key and see how Stellarium returns to the original field of view (how "zoomed in" the view is), and direction of view.

It's also possible to move around using the mouse. If you left-click and drag somewhere on the sky, you can pull the view around.

Another method of moving is to select some object in the sky (left-click on the object), and press the Space key to centre the view on that object. Similarly, selecting an object and pressing the forward-slash key / will centre on the object and zoom right in on it.

The forward-slash / and backslash \ keys auto-zoom in an out to different zoom levels depending on what is selected. If the object selected is a planet or moon in a *sub-system* with a lot of moons (e.g. Jupiter), the initial zoom in will go to an intermediate level where the whole sub-system should be visible. A second zoom will go to the full zoom level on the selected object. Similarly, if you are fully zoomed in on a moon of Jupiter, the first auto-zoom out will go to the sub-system zoom level. Subsequent auto-zoom out will fully zoom out and return the initial direction of view. For objects that are not part of a sub-system, the initial auto-zoom in will zoom right in on the selected object (the exact field of view depending on the size/type of the selected object), and the initial auto-zoom out will return to the initial FOV and direction of view.

3.3   The Main Tool Bar

Figure 3.2: Night scene with constellation artwork and moon.

Stellarium can do a whole lot more than just draw the stars. Figure 3.2 shows some of Stellarium's visual effects including constellation line and boundary drawing, constellation art, planet hints, and atmospheric halo around the bright Moon. The controls in the main tool-bar provide a mechanism for turning on and off the visual effects.

When the mouse if moved to the bottom left of the screen, a second tool-bar becomes visible. All the buttons in this side tool-bar open and close dialog boxes which contain controls for further configuration of the program. The dialogs will be described in the next chapter.

Table 3.3 describes the operations of buttons on the main tool-bar and the side tool-bar, and gives their default keyboard shortcuts.

Feature	Button	Key	Description
Constellations		C	Draw constellations as "stick figures"
Constellation Names		V	Draw name of the constellations

Constellation Art		R	Superimpose artistic representations of the constellations
Equatorial Grid		E	Draw grid lines for the RA/Dec coordinate system
Azimuth Grid		Z	Draw grid lines for the Alt/Azi coordinate system
Toggle Ground		G	Toggle drawing of the ground. Turn this off to see objects that are below the horizon.
Toggle Cardinal Points		Q	Toggle marking of the North, South, East and West points on the horizon.
Toggle Atmosphere		A	Toggle atmospheric effects. Most notably makes the stars visible in the daytime.
Deep-Sky Objects		D	Toggle marking the positions of Deep-Sky Objects.
Planet Hints		P	Toggle indicators to show the position of planets.
Coordinate System		Ctrl + M	Toggle between Alt/Azi & RA/Dec coordinate systems.
Goto			Center the view on the selected object
Night Mode		Ctrl + N	Toggle "night mode", which applies a red-only filter to the view to be easier on the dark-adapted eye.
Nebula images		I	Toggle "nebula images". This button must be enabled first, see section 4.3.4
Full Screen Mode		F11	Toggle full screen mode.
Flip view (horizontal)		Ctrl + Shift + H	Flip the image in the horizontal plane. This button must be enabled first, see section 4.3.4
Flip view (vertical)		Ctrl + Shift + V	Flip the image in the vertical plane. This button must be enabled first, see section 4.3.4
Quit Stellarium		Ctrl + Q	Close Stellarium.
Help Window		F1	Show the help window, with key bindings and other useful information

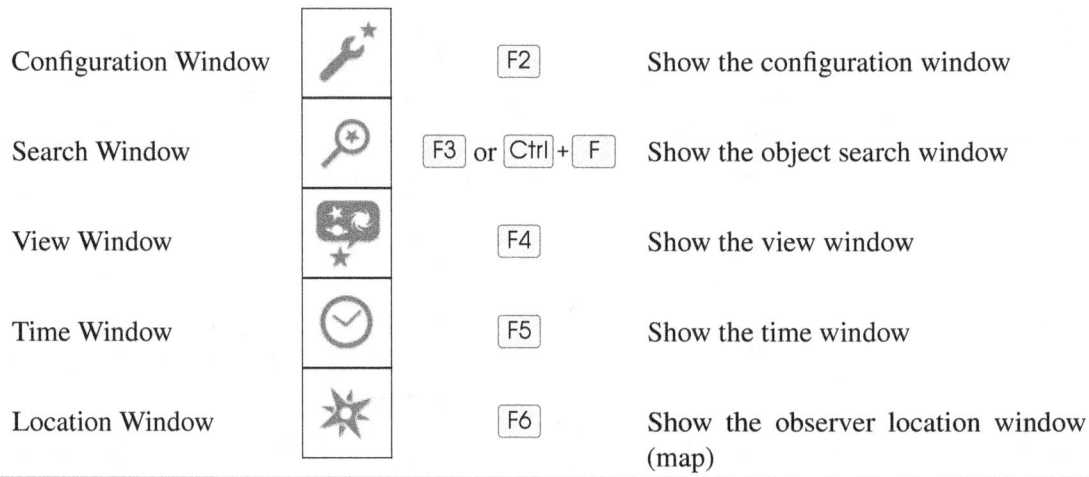

Configuration Window		F2	Show the configuration window
Search Window		F3 or Ctrl + F	Show the object search window
View Window		F4	Show the view window
Time Window		F5	Show the time window
Location Window		F6	Show the observer location window (map)

Table 3.3: Stellarium's standard menu buttons

3.4 Taking Screenshots

You can save what is on the screen to a file by pressing Ctrl + S . Screenshots are taken in .png format, and have filenames like stellarium-000.png, stellarium-001.png (the number increments to prevent overwriting existing files).

Stellarium creates screenshots in a directory depending on your operating system, see section 5.1 Files and Directories.

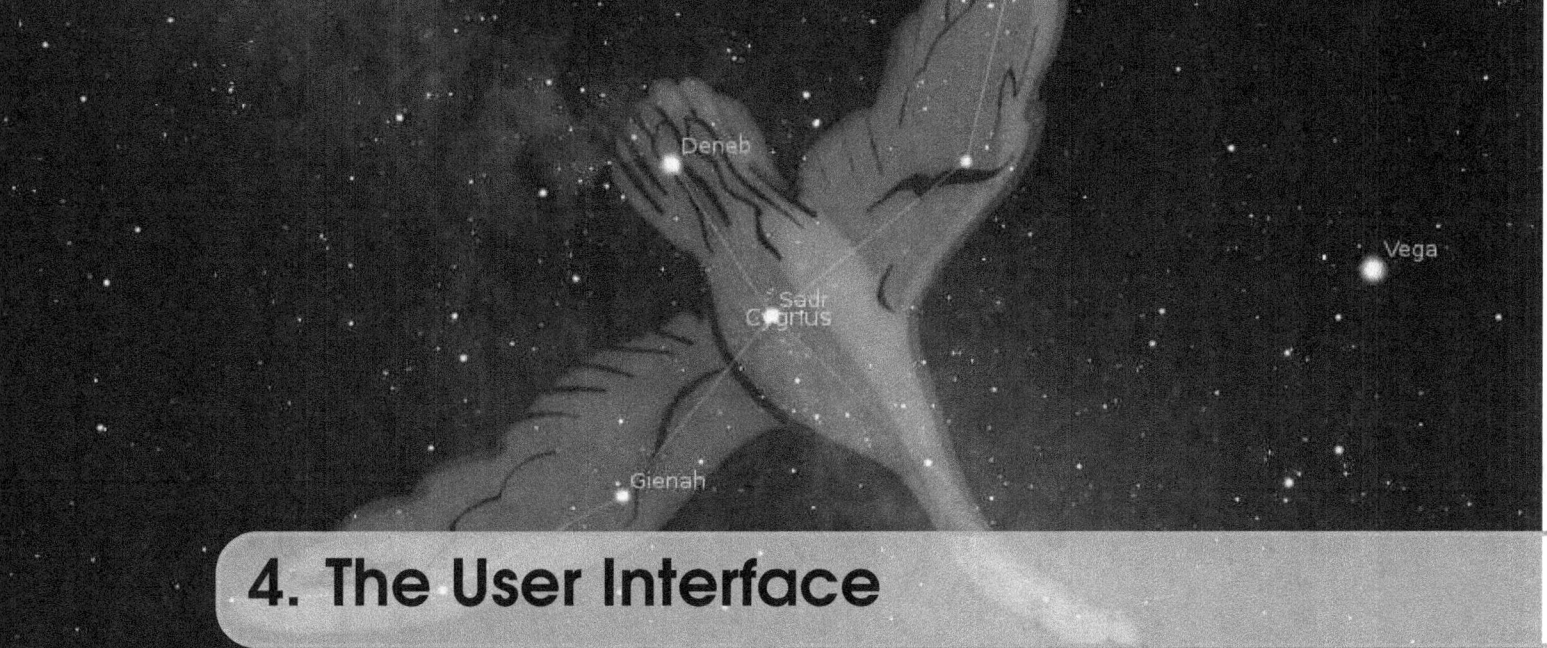

4. The User Interface

This chapter describes the dialog windows which can be accessed from the left menu bar.

Most of Stellarium's settings can be changed using the view window (press ![icon] or F4) and

the configuration window (![icon] or F2). Most settings have short labels. To learn more about some settings, more information is available as *tooltips*, small text boxes which appear when you hover the mouse cursor over a button.[1]

0.15

You can drag the windows around, and the position will be used again when you restart Stellarium. If this would mean the window is off-screen (because you start in windowed mode, or with a different screen), the window will be moved so that at least a part is visible.

Some options are really rarely changed and therefore may only be configured by editing the configuration file. See 5.3 The Main Configuration File for more details.

4.1 Setting the Date and Time

Figure 4.1: Date and Time dialog

In addition to the time rate control buttons on the main toolbar, you can use the date and time

---

[1]Unfortunately, on Windows 7 and later, with NVidia and AMD GPUs, these tooltips often do not work.

window (open with the 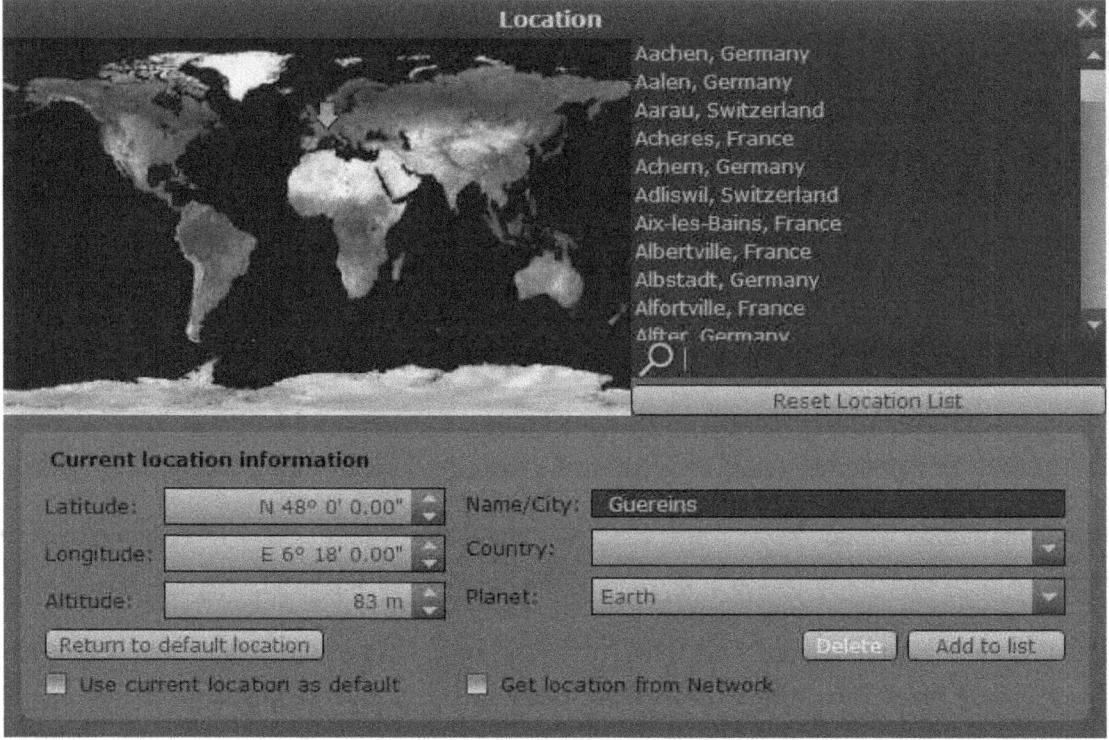 button or F5) to set the simulation time. The values for year, month, day, hour, minutes and seconds may be modified by typing new values, by clicking the up and down arrows above and below the values, and by using the mouse wheel.

The other tab in this window allows you to see or set *Julian Day* and/or *Modified Julian Day* numbers (see 16.3.3).

4.2 Setting Your Location

Figure 4.2: Location window

The positions of the stars in the sky is dependent on your location on Earth (or other planet) as well as the time and date. For Stellarium to show accurately what is (or will be/was) in the sky, you must tell it where you are. You only need to do this once – Stellarium can save your location so you won't need to set it again until you move.

After installation, Stellarium uses an online service which tries to find your approximate location based on the IP address you are using. This seems very practical, but if you feel this causes privacy issues, you may want to switch this feature off. You should also consider switching it off on a computer which does not move, to save network bandwidth.

To set your location more accurately, or if the lookup service fails, press F6 to open the location window. There are a few ways you can set your location:

1. Just click on the map.
2. Search for a city where you live using the search edit box at the top right of the window, and select the right city from the list.
3. Click on the map to filter the list of cities in the vicinity of your click, then choose from the shortlist.
4. Enter a new location using the longitude, latitude and other data.

If you want to use this location permanently, click on the "use as default" checkbox, disable "Get location from Network", and close the location window.

4.3 The Configuration Window

The configuration window contains general program settings, and many other settings which do not concern specific display options. Press the tool button [🔧] or [F2] to open.

4.3.1 The Main Tab

The Main tab in the configuration window provides controls for changing separately the program and sky culture languages.

The next setting group allows to enable using DE430/DE431 ephemeris files. Most users do not require this. Thes files have to be installed separately. See section 5.4.1 if you are interested.

The tab also provides a button for saving the current program configuration. Most display settings have to be explicitly stored to make a setting change permanent.

4.3.2 The Information Tab

The Information tab allows you to set the type and amount of information displayed about a selected object.

- Ticking or unticking the relevant boxes will control this.
- The information displays in various colours depending on the type and level of the stored data

4.3.3 The Navigation Tab

The Navigation tab (Fig. 4.5) allows for enabling/disabling of keyboard shortcuts for panning and zooming the main view, and also how to specify what simulation time should be used when the program starts:

System date and time Stellarium will start with the simulation time equal to the operating system clock.

System date at Stellarium will start with the same date as the operating system clock, but the time will be fixed at the specified value. This is a useful setting for those people who use Stellarium during the day to plan observing sessions for the upcoming evening.

Other some fixed time can be chosen which will be used every time Stellarium starts.

The lowest field allows selection of the correction model for the time correction ΔT (see section 16.3.3). Default is "Espenak and Meeus (2006)". Please use other values only if you know what you are doing.

4.3.4 The Tools Tab

The Tools tab (Fig. 4.6) contains miscellaneous utility features:

Spheric mirror distortion This option pre-warps the main view such that it may be projected onto a spherical mirror using a projector. The resulting image will be refected up from the spherical mirror in such a way that it may be shone onto a small planetarium dome, making a cheap planetarium projection system.

Select single constellation When active, clicking on a star that is member in the constellation lines will make the constellation stand out. You can select several constellations, but clicking onto a star which is not member of a constellation line will display all constellations.

Show nebula background button You can disable display of DSO photographs with this button.

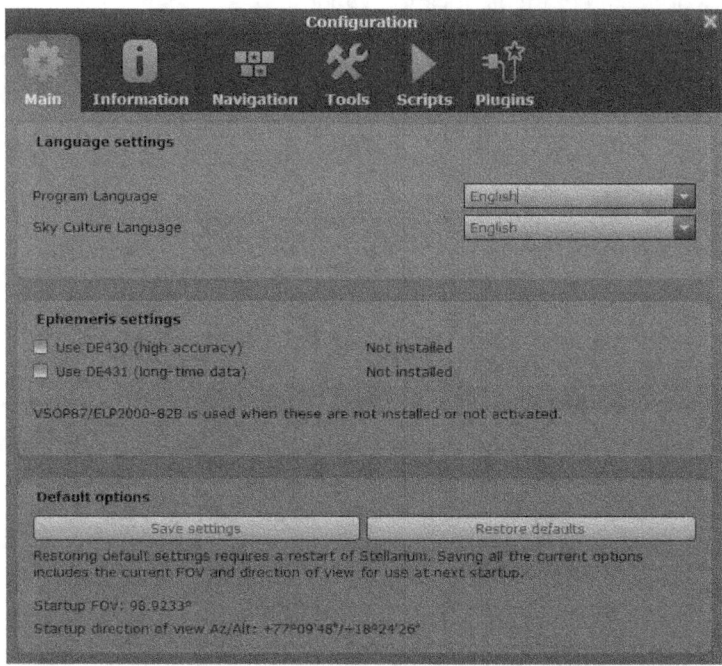

Figure 4.3: Configuration Window: Main Tab

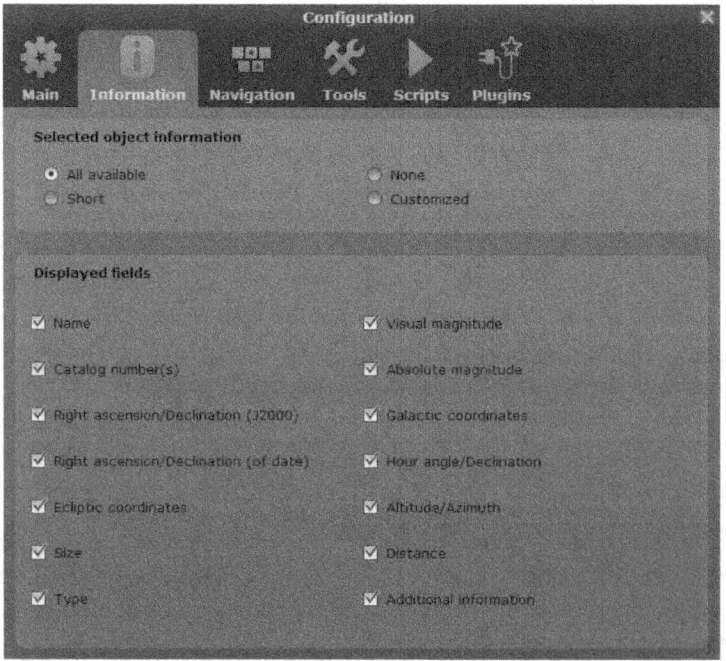

Figure 4.4: Configuration Window: Information Tab

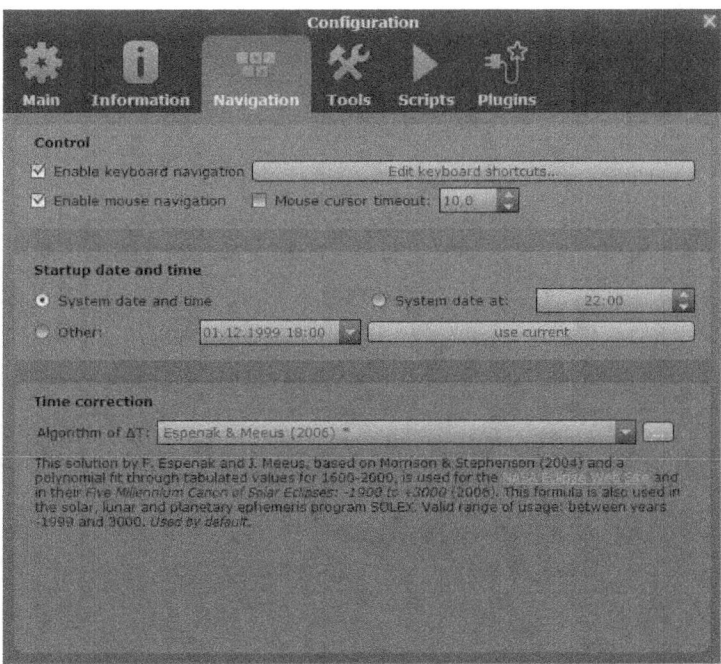

Figure 4.5: Configuration Window: Navigation Tab

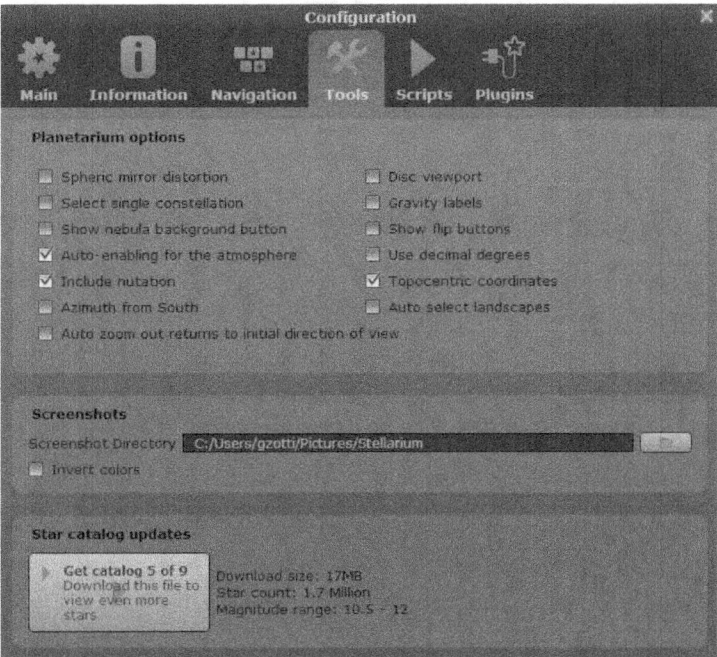

Figure 4.6: Configuration Window: Tools Tab

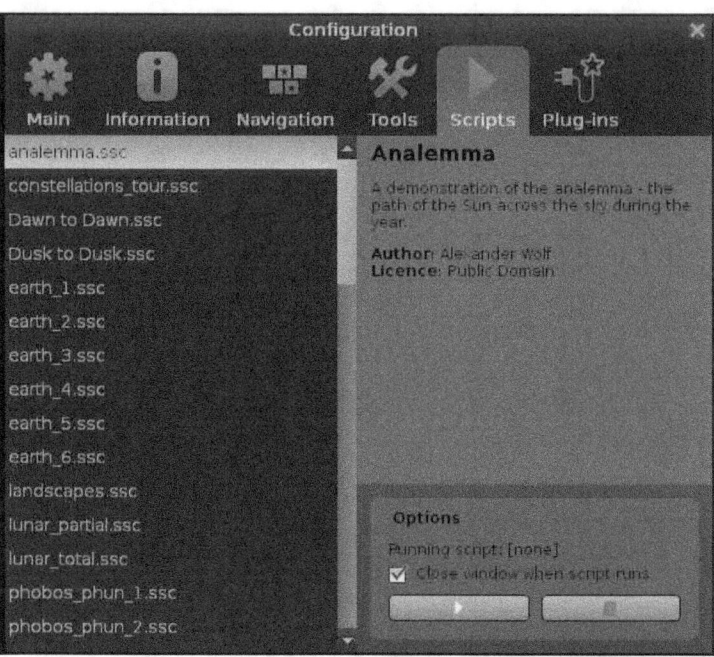

Figure 4.7: Configuration Window: Scripts Tab

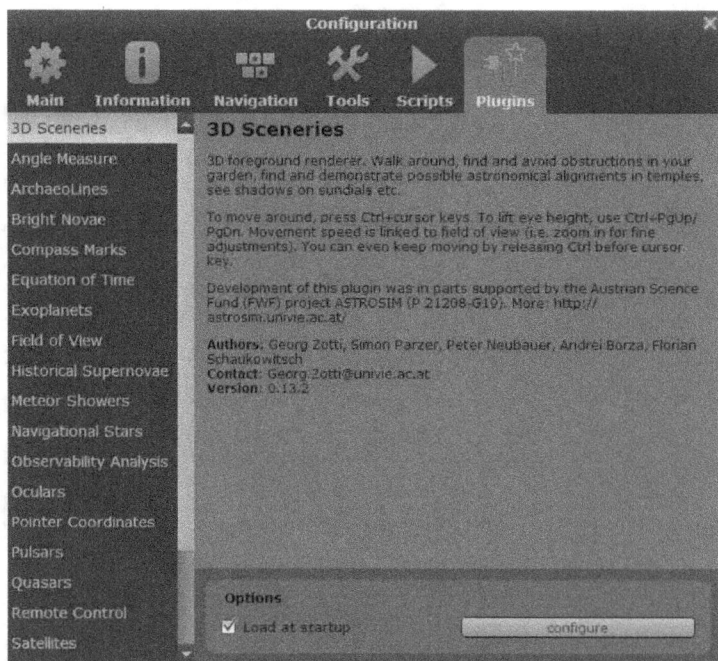

Figure 4.8: Configuration Window: Plugins Tab

Auto-enabling for the atmosphere When changing planet during location change, atmosphere will be switched as required.

Include nutation Compute the slight wobble of earth's axis. This feature is active only about 500 years around J2000.0.

Azimuth from South Some users may be used to counting azimuth from south.

Disc viewport This option masks the main view producing the effect of a telescope eyepiece. It is also useful when projecting Stellarium's output with a fish-eye lens planetarium projector.

Gravity labels This option makes labels of objects in the main view align with the nearest horizon. This means that labels projected onto a dome are always aligned properly.

Show flip buttons When enabled, two buttons will be added to the main tool bar which allow the main view to be mirrored in the vertical and horizontal directions. This is useful when observing through telecopes which may cause the image to be mirrored.

Use decimal degrees

Topocentric coordinates If you require planetocentric coordinates, you may switch this off. Usually it should be enabled.

Auto select landscapes When changing the planet in the location panel, a fitting landscape panorama will be shown when available.

Auto zoom out returns to initial field of view When enabled, this option changes the behaviour of the zoom out key (\) so that it resets the initial direction of view in addition to the field of view.

4.3.5 The Scripts Tab

The Scripts tab (Fig. 4.7) allows the selection of pre-assembled scripts bundled with Stellarium that can be run (See chapter 15 for an introduction to the scripting capabilities and language). This list can be expanded by your own scripts as required. See section 5.2 where to store your own scripts.

When a script is selected it can be run by pressing the arrow button and stopped with the stop button. With some scripts the stop button is inhibited until the script is finished.

Scripts that use sound or embedded videos will need a version of Stellarium configured at compile time with multimedia support enabled. It must be pointed out here that sound or video codecs available depends on the sound and video capabilities of you computer platform and may not work.

4.3.6 The Plugins Tab

Plugins (see chapter 10 for an introduction) can be enabled here (Fig. 4.8) to be loaded the next time you start Stellarium. When loaded, many plugins allow additional configuration which is available by pressing the configure button on this tab.

4.4 The View Settings Window

The View settings window controls many display features of Stellarium which are not available via the main toolbar.

4.4.1 Sky Tab

The Sky tab of the View window (Fig. 4.9) contains settings for changing the general appearance of the main sky view. Some hightlights:

Absolute scale is the size of stars as rendered by Stellarium. If you increase this value, all stars will appear larger than before.

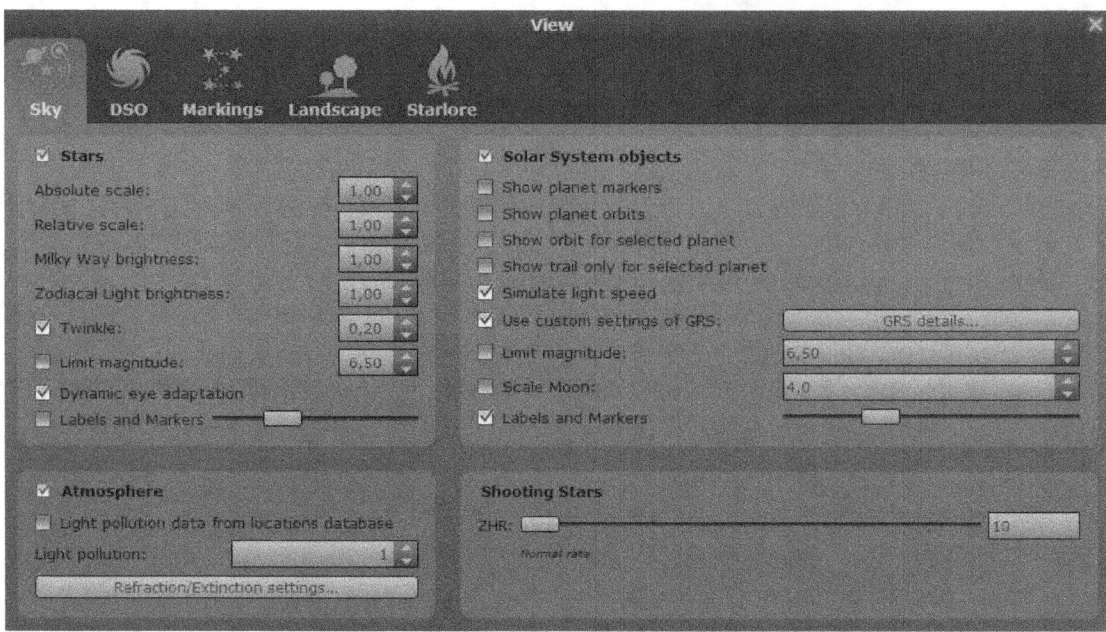

Figure 4.9: View Settings Window: Sky Tab

Relative scale determines the difference in size of bright stars compared to faint stars. Values higher than 1.00 will make the brightest stars appear much larger than they do in the sky. This is useful for creating star charts, or when learning the basic constellations.

Twinkle controls how much the stars twinkle when atmosphere is enabled. Since V0.15, the twinkling is reduced in higher altitudes, where the star light passes the atmosphere in a steeper angle and is less distorted.

Limit magnitude Inhibits automatic addition of fainter stars when zooming in. This may be helpful if you are interested in naked eye stars only.

Dynamic eye adaptation When enabled this feature reduces the brightness of faint objects when a bright object is in the field of view. This simulates how the eye can be dazzled by a bright object such as the moon, making it harder to see faint stars and galaxies.

Light pollution In urban and suburban areas, the sky is brightned by terrestrial light pollution reflected in the atmophere. Stellarium simulates light pollution and is calibrated to the *Bortle Dark Sky Scale* where 1 means a good dark sky, and 9 is a very badly light-polluted sky. See Appendix B for more information.

Solar System objects this group of options lets you turn on and off various features related to the planets. Simulation of light speed will give more precise positions for planetary bodies which move rapidly against backround stars (e.g. the moons of Jupiter). The *Scale Moon* option will increase the apparent size of the moon in the sky, which can be nice for wide field of view shots.

Labels and markers you can independantly change the amount of labels displayed for planets, stars and nebuulae. The further to the right the sliders are set, the more labels you will see. Note that more labels will also appear as you zoom in.

Shooting stars Stellarium has a simple meteor simulation option. This setting controls how many shooting stars will be shown. Note that shooting stars are only visible when the time rate is 1, and might not be visiable at some times of day. Meteor showers are not currently simulated.

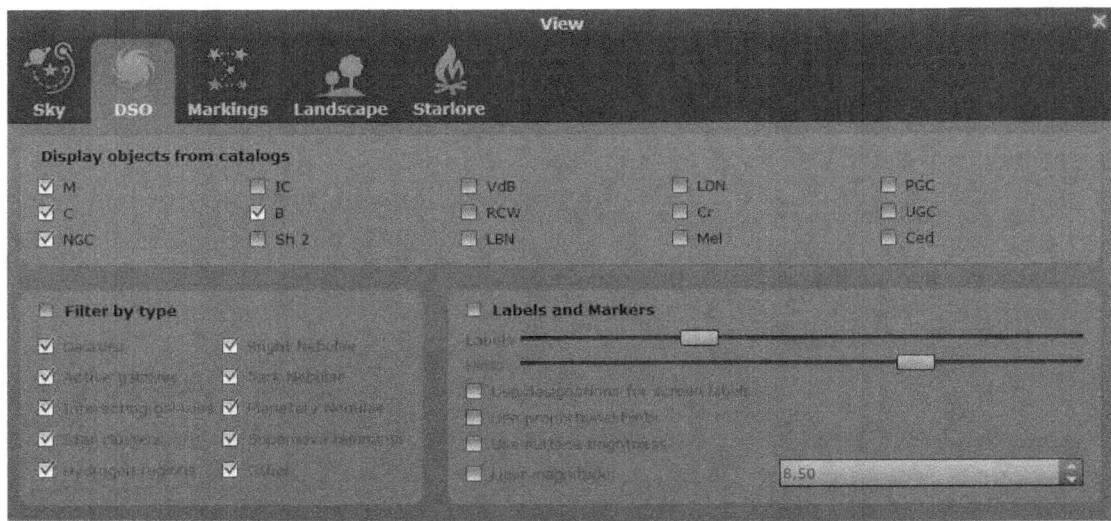

Figure 4.10: View Settings Window: DSO Tab

Atmosphere settings

An auxiliary dialog contains detail settings for the atmosphere. Here you can set atmospheric pressure and temperature which influence refraction (see section 17.13.1), and the opacity factor for extinction, *magnitude loss per airmass k* (see section 17.13.2).

4.4.2 DSO Tab

Deep-sky objects or DSO are extended objects which are external to the solar system, and are not point-sources like stars. DSO include galaxies, planetary nebulae and star clusters. These objects may or may not have images associated with them. Stellarium comes with a catalogue with over 14,000 extended objects containing the combined data from many catalogues, with 190 images. The DSO tab (Fig. 4.10) allows you to specify which catalogs or which object types you are interested in. See chapter 8 for details about the catalog, and how to extend it with your own photographs.

4.4.3 Markings Tab

The Markings tab of the View window (Fig. 4.11) controls the following features:

Celestial sphere this group of options makes it possible to plot various grids and lines in the main view.

Projection Selecting items in this list changes the projection method which Stellarium uses to draw the sky [57]. Options are:

    Perspective Perspective projection maps the horizon and other great circles like equator, ecliptic, hour lines, etc. into straight lines. The maximum field of view is 150°. The mathematical name for this projection method is *gnomonic projection*.

    Stereographic Stereographic projection has been known since antiquity and was originally known as the planisphere projection. It preserves the angles at which curves cross each other but it does not preserve area. Else it is similar to fish-eye projection mode. The maximum field of view in this mode is 235°.

    Fish-Eye Stellarium draws the sky using *azimuthal equidistant projection*. In fish-eye projection, straight lines become curves when they appear a large angular distance from the centre of the field of view (like the distortions seen with very wide angle camera lenses). This is more pronounced as the user zooms out. The maximum field of view in this mode is 180°.

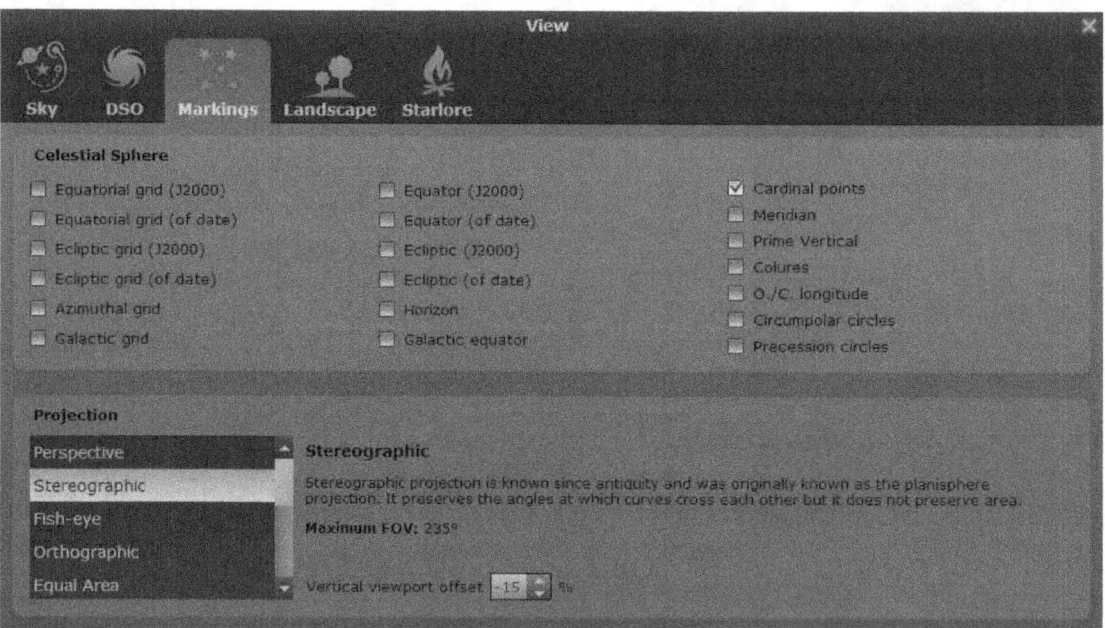

Figure 4.11: View Settings Window: Markings Tab

Orthographic Orthographic projection is related to perspective projection, but the *point of perspective* is set to an infinite distance. The maximum field of view is 180°.

Equal Area The full name of this projection method is *Lambert azimuthal equal-area projection*. It preserves the area but not the angle. The maximum field of view is 360°.

Hammer-Aitoff The Hammer projection is an equal-area map projection, described by ERNST HAMMER in 1892 and directly inspired by the Aitoff projection. The maximum field of view in this mode is 360°.

Sinusoidal The sinusoidal projection is a *pseudocylindrical equal-area map projection*, sometimes called the Sanson–Flamsteed or the Mercator equal-area projection. Meridians are mapped to sine curves.

Mercator Mercator projection is a cylindrical projection which preserves the angles between objects, and the scale around an object is the same in all directions. The poles are mapped to infinity. The maximum field of view in this mode is 233°.

Miller cylindrical The Miller cylindrical projection is a modified Mercator projection, proposed by OSBORN MAITLAND MILLER (1897–1979) in 1942. The poles are no longer mapped to infinity.

Cylinder The full name of this simple projection mode is *cylindrical equidistant projection* or *Plate Carrée*. The maximum field of view in this mode is 233°.

4.4.4 Landscape Tab

The Landscape tab of the View window (Fig. 4.12) controls the landscape graphics (the horizon which surrounds you). To change the landscape graphics, select a landscape from the list on the left side of the window. A description of the landscape will be shown on the right.

Note that while a landscape can include information about where the landscape graphics were taken (planet, longitude, latitude and altitude), this location does not have to be the same as the location selected in the Location window, although you can set up Stellarium such that selection of a new landscape will alter the location for you.

The controls at the bottom right of the window operate as follows:

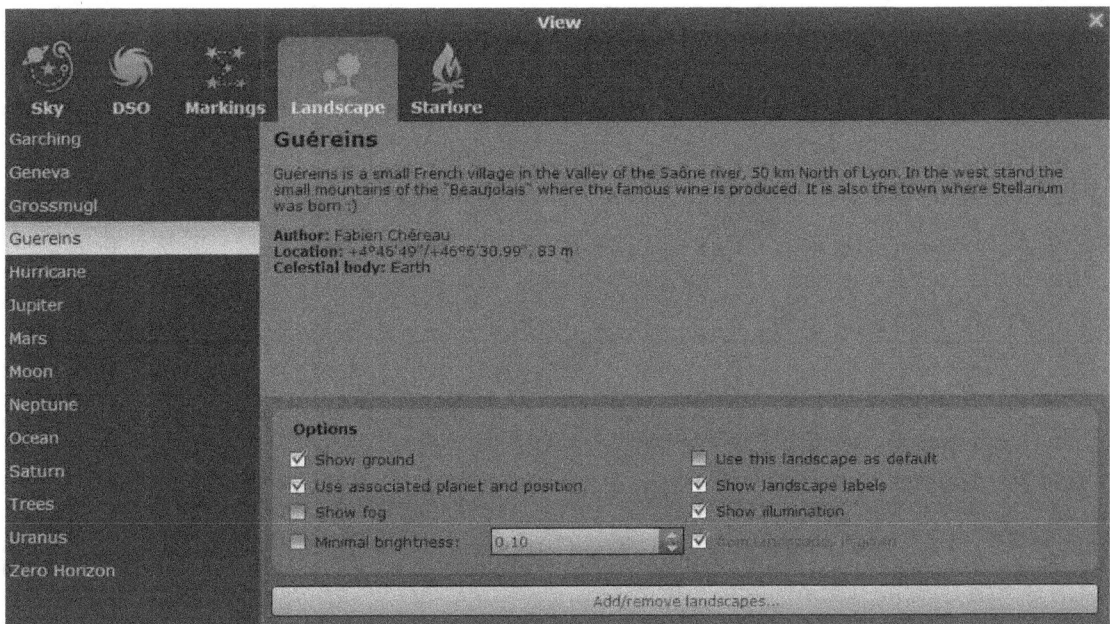

Figure 4.12: View Settings Window: Landscape Tab

Show ground This turns on and off landscape rendering (same as the button in the main tool-bar).

Show_fog This turns on and off rendering of a band of fog/haze along the horizon, when available in this landscape.

Use associated planet and position When enabled, selecting a new landscape will automatically update the observer location.

Use this landscape as default Selecting this option will save the landscape into the program configuration file so that the current landscape will be the one used when Stellarium starts.

Minimal brightness Use some minimal brightness setting. Moonless night on very dark locations may appear too dark on your screen. You may want to configure some minimal brightness here.

from landscape, if given Landscape authors may decide to provide such a minimal brightness value in the landscape.ini file.

Show landscape labels Landscapes can be configured with a gazetteer of interesting points, e.g., mountain peaks, which can be labeled with this option.

Show illumination to reflect the ugly developments of our civilisation, landscapes can be configured with a layer of light pollution, e.g., streetlamps, bright windows, or the sky glow of a nearby city. This layer, if present, will be mixed in when it is dark enough.

Using the button Add/remove landscapes... , you can also install new landscapes from ZIP files which you can download e.g. from the Stellarium website[2] or create yourself (see ch. 7 Landscapes), or remove these custom landscapes.

4.4.5 Starlore Tab

The Starlore tab of the View window (Fig. 4.13) controls what culture's constellations and bright star names will be used in the main display. Some cultures have constellation art (e.g., Western and Inuit), and the rest do not. Configurable options include

---

[2]http://stellarium.org/wiki/index.php/Landscapes

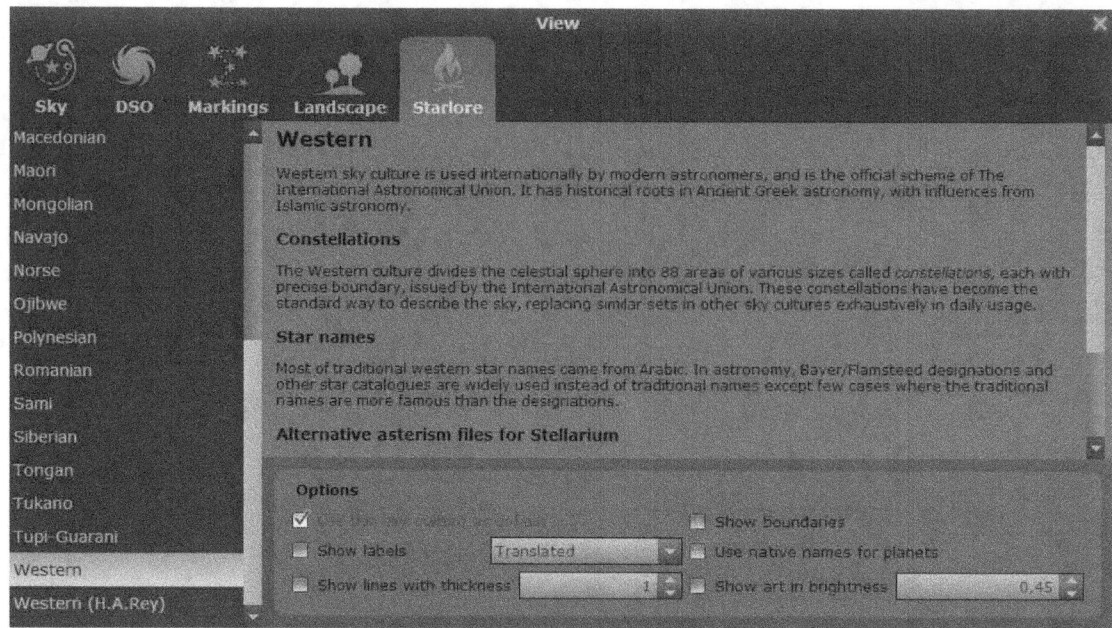

Figure 4.13: View Settings Window: Starlore Tab

Use this skyculture as default Activate this option to load this skyculture when Stellarium starts.

Show labels Activate display of constellation labels, like $\boxed{\mathcal{T}}$ or $\boxed{\text{V}}$. You can further select whether you want to display abbreviated, original or translated names.

Show lines with thickness... Activate display of stick figures, like $\boxed{\mathcal{N}}$ or $\boxed{\text{C}}$, and you can configure line thickness here.

Show boundaries Activate display of constellation boundaries, like $\boxed{\text{B}}$. Currently, boundaries have been defined only for "Western" skycultures.

Use native names for planets If provided, show the planet names as used in this skyculture (also shows modern planet name for reference).

Show art in brightness... Activate display of constellation art (if available), like $\boxed{\mathcal{K}}$ or $\boxed{\text{R}}$. You can also select the brightness here.

4.5   The Object Search Window

The Object Search window provides a convenient way to locate objects in the sky. Simply type in the name of an object to find, and press $\boxed{\downarrow}$. Stellarium will point you at that object in the sky.

As you type, Stellarium will make a list of objects which contains what you have typed so far. The first of the list of matching objects will be highlighted. If you press the $\boxed{\hookrightarrow}$ key, the selection will change to the next item in the list. Hitting the $\boxed{\downarrow}$ key will go to the currently highlighted object and close the search dialog.

For example, suppose we want to locate Mimas (a moon of Saturn). After typing the first letter of the name, *m*, Stellarium makes a list of objects whose name contains M: Haumea, Miranda, Umbriel, . . .

You may want at this point to have Stellarium rather propose object names with start with the string you enter. Do that in the Options tab of this panel. Now repeat searching (delete, and re-enter M to start over). Now the list is shorter and contains only objects which start with M: Maia, Mars, . . . The first item in this list, Maia, is highlighted. Pressing $\boxed{\downarrow}$ now would go to Maia, but we

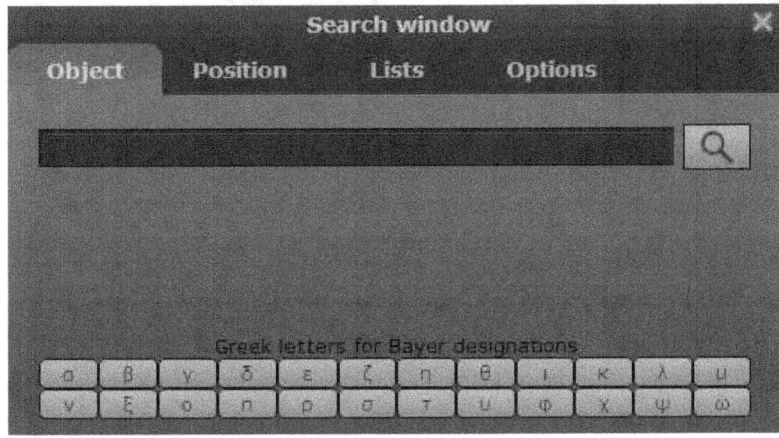

Figure 4.14: The Search Window: Objects

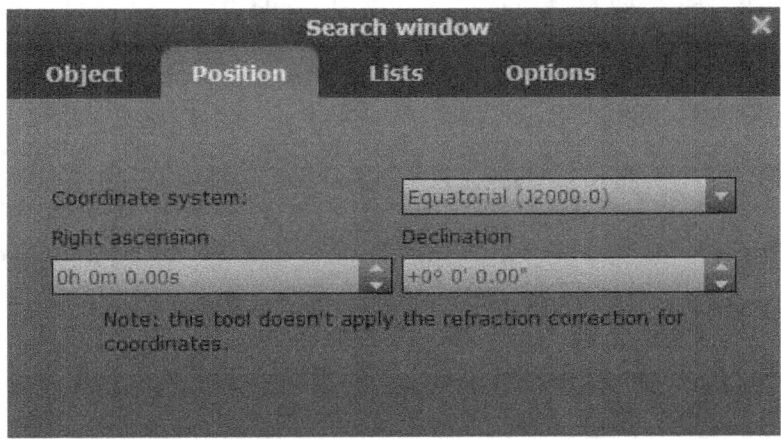

Figure 4.15: The Search Window: Positions

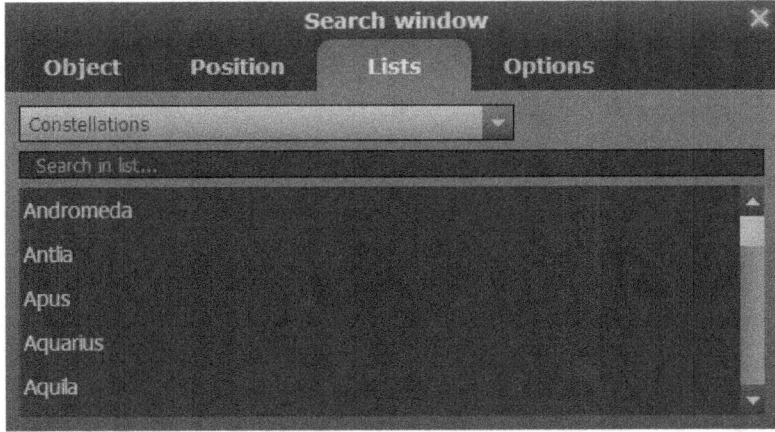

Figure 4.16: The Search Window: Lists

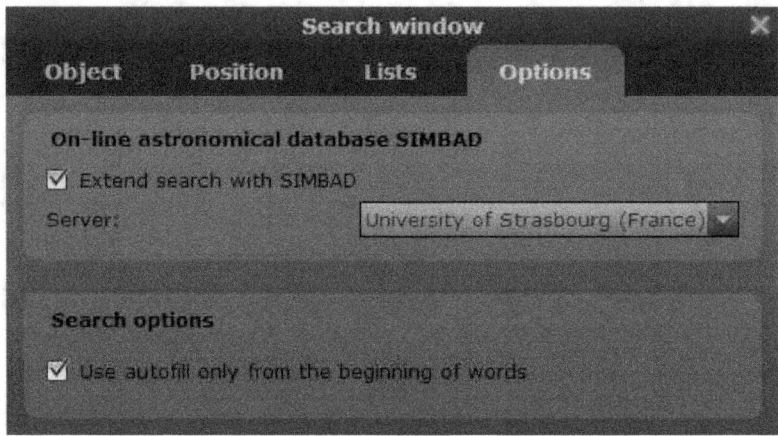

Figure 4.17: The Search Window: Options

want Mimas. We can either press ⬚ a few times to highlight Mimas and then hit ⏎, or we can continue to type the name until it is the first/only object in the list.

The Position tab provides a convenient way to enter a set of coordinates.

The List Search tab allows selection of an object from predefined sets. The number of choices is governed by the loaded plug ins. Simply scroll down the first window to select the type. The name of an object can then be selected from the list. Press ⏎ and Stellarium will go to that object.

The Options tab provides a few settings to fine-tune your search experience. When the name of an object to find is typed in the object window and you are connected to the internet and "Extend search" is ticked, Stellarium will search the SIMBAD on-line data bases for its coordinates. You can then click the go button or press return. Stellarium will point you at that object in the sky even if there is no object displayed on the screen. The SIMBAD server being used can be selected from the scroll window.

4.6   The AstroCalc Window

This window provides advanced functionality and is still in an experimental phase. You can call it by pressing F10.

The AstroCalc window shows three tabs with different functionality.

Planet Positions Shows J2000.0 positions and magnitudes for all installed planets, planet moons, minor bodies (asteroids, comets, etc.). Clicking on an entry brings the object into focus.

Ephemeris Select an object and start and end time, and compute an ephemeris (list of positions and magnitude evolving over time) for that object. The positions are marked in the sky with yellow circles (Figure 4.18). When you click on a date, an orange circle indicates this date. Double-clicking sets the respective date and brings object to focus. You can export the calculated ephemeris into CSV file.

Phenomena Compute phenomena like conjunctions and oppositions between planetary objects. You can export the calculated conjunctions and oppositions into CSV file.

4.7   Help Window

The Help window lists all of Stellarium's keystrokes. Note that some features are only available as keystrokes, so it's a good idea to have a browse of the information in this window.

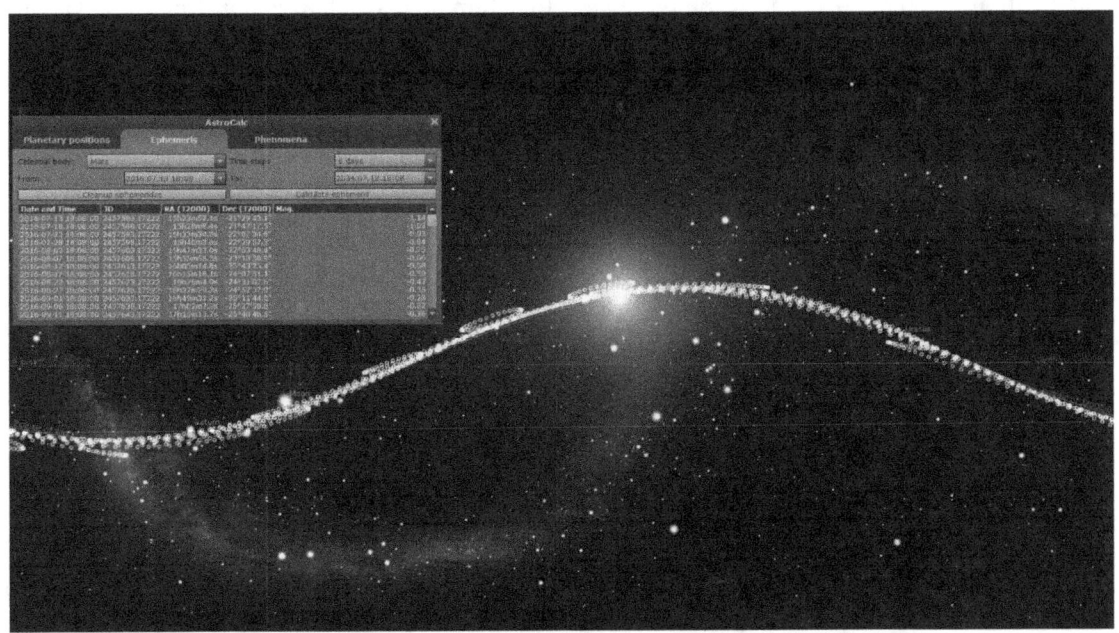

Figure 4.18: AstroCalc: Plot traces of planets.

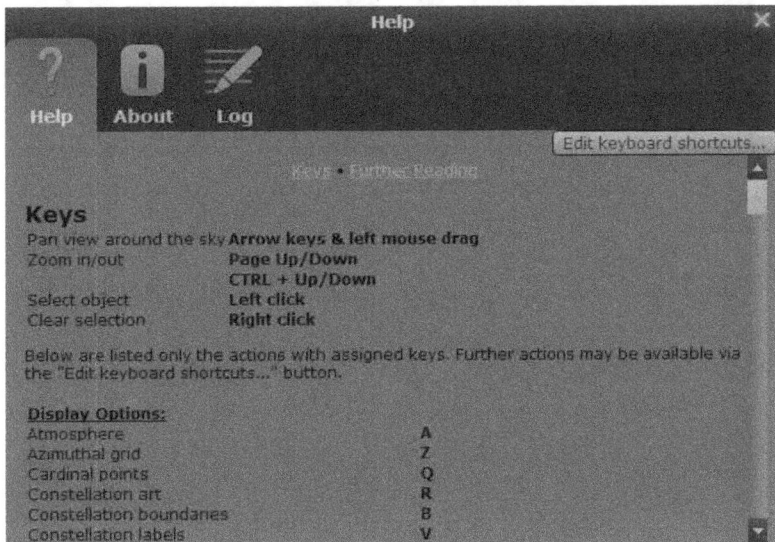

Figure 4.19: Help Window

4.7.1 Editing Keyboard Shortcuts

You can edit the shortcut keys here. Each available function can be configured with up to two key combinations. You may want to reconfigure keys for example if you have a non-English keyboard layout and some keys either do not work at all, or feel unintuitive for you, or if you are familiar with other software and want to use the same hotkeys for similar functions. Simply select the function and click with the mouse into the edit field, then press your key of choice. If the key has been taken already, a message will tell you.

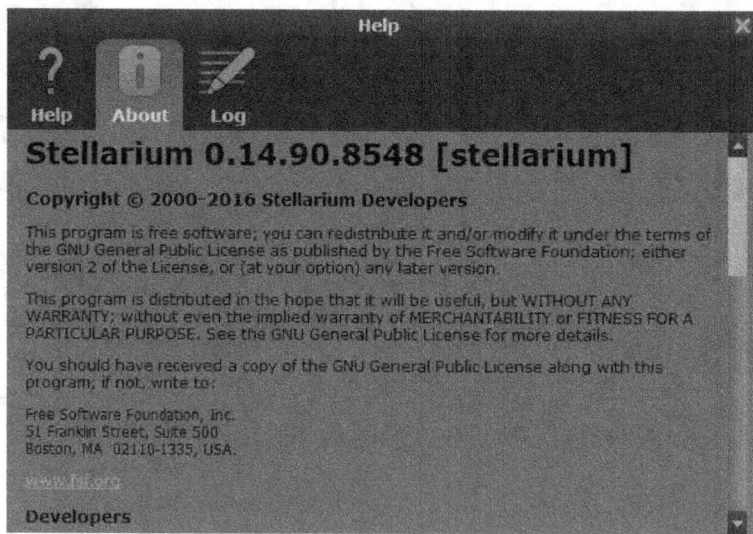

Figure 4.20: Help Window: About

The About tab (Fig. 4.20) shows version and licensing information, and a list of people who helped to produce the program.

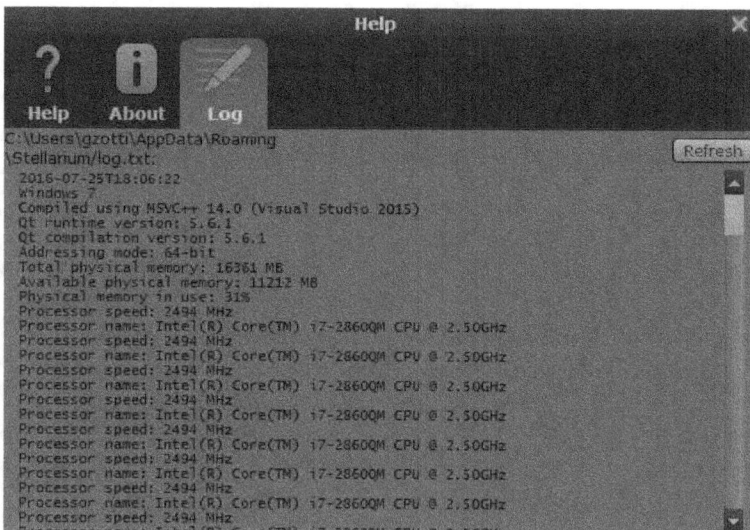

Figure 4.21: Help Window: Logfile

The log tab (Fig. 4.21) shows messages like the loading confirmations carried out when stellarium runs. It is useful to locate the files that stellarium writes to your computer. The same information is written to the file log.txt that you will find in your user directory (see 5.1).

5 Files and Directories 49
5.1 Directories
5.2 Directory Structure
5.3 The Main Configuration File
5.4 Getting Extra Data

6 Command Line Options 53
6.1 Examples

7 Landscapes 57
7.1 Stellarium Landscapes
7.2 Creating Panorama Photographs for Stellarium
7.3 Panorama Postprocessing
7.4 Other recommended software

8 Deep-Sky Objects 81
8.1 Stellarium DSO Catalog
8.2 Adding Extra Nebulae Images

9 Adding Sky Cultures 89
9.1 Basic Information
9.2 Skyculture Description Files
9.3 Constellation Names
9.4 Star Names
9.5 Planet Names
9.6 Stick Figures
9.7 Constellation Borders
9.8 Constellation Artwork
9.9 Seasonal Rules
9.10 Publish Your Work

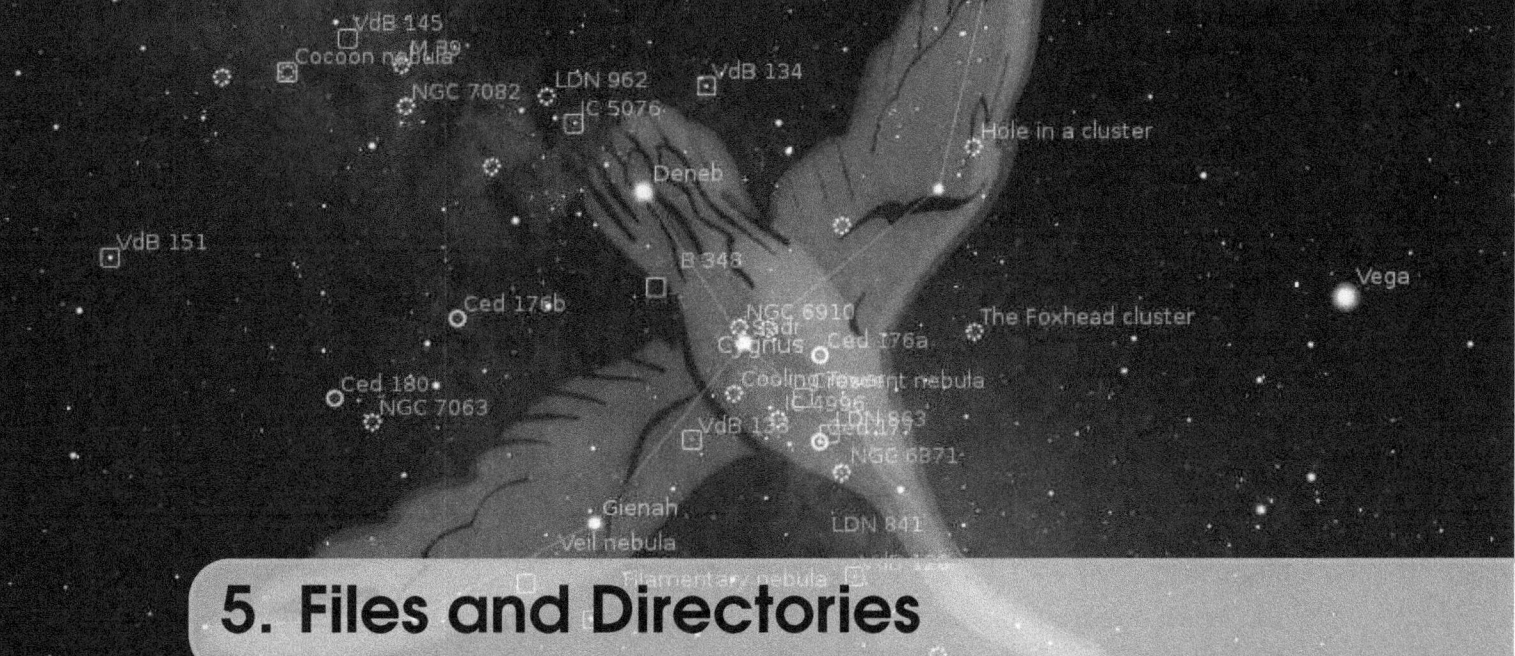

5. Files and Directories

5.1 Directories

Stellarium has many data files containing such things as star catalogue data, nebula images, button icons, font files and configuration files. When Stellarium looks for a file, it looks in two places. First, it looks in the *user directory* for the account which is running Stellarium. If the file is not found there, Stellarium looks in the *installation directory*[1]. Thus it is possible for Stellarium to be installed by an administrative user and yet have a writable configuration file for non-administrative users. Another benefit of this method is on multi-user systems: Stellarium can be installed by the administrator, and different users can maintain their own configuration and other files in their personal user accounts.

In addition to the main search path, Stellarium saves some files in other locations, for example screens shots and recorded scripts.

The locations of the user directory, installation directory, *screenshot save directory* and *script save directory* vary according to the operating system and installation options used. The following sections describe the locations for various operating systems.

5.1.1 Windows

installation directory By default this is `C:\Program Files\Stellarium\`, although this can be adjusted during the installation process.

user directory This is the Stellarium sub-folder in the Application Data folder for the user account which is used to run Stellarium. Depending on the version of Windows and its configuration, this could be any of the following (each of these is tried, if it fails, the next in the list if tried).

```
%APPDATA%\Stellarium\
%USERPROFILE%\Stellarium\
%HOMEDRIVE%\%HOMEPATH%\Stellarium\
%HOME%\Stellarium\
Stellarium's installation directory
```

---

[1]The installation directory was referred to as the config root directory in previous versions of this guide

Thus, on a typical Windows Vista/7/10 system with user "Bob Dobbs", the user directory will be:

```
C:\Users\Bob Dobbs\AppData\Roaming\Stellarium\
```

The user data directory is unfortunately hidden by default. To make it accessible in the Windows file explorer, open an **Explorer** window and select Organize... ⟩ Folder and search options . Make sure folders marked as hidden are now displayed. Also, deselect the checkbox to "hide known file name endings".[2]

screenshot save directory Screenshots will be saved to the `Pictures/Stellarium` directory, although this can be changed with a command line option (see section 6[3]).

5.1.2 Mac OS X

installation directory This is found inside the application bundle, `Stellarium.app`. See *Inside Application Bundles*[4] for more information.

user directory This is the sub-directory `Library/Preferences/Stellarium/` (or `~/Library/Application Support/Stellarium` on newest versions of Mac OS X) of the user's home directory.

screenshot save directory Screenshots are saved to the user's Desktop.

5.1.3 Linux

installation directory This is in the `share/stellarium` sub-directory of the installation prefix, i.e., usually `/usr/share/stellarium` or `/usr/local/share/stellarium/`.

user directory This is the `.stellarium` sub-directory of user's home directory, i.e., `~/.stellarium/`. This is a hidden folder, so if you are using a graphical file browser, you may want to change its settings to "display hidden folders".

screenshot save directory Screenshots are saved to the user's home directory.

5.2 Directory Structure

Within the *installation directory* and *user directory* defined in section 5.1, files are arranged in the following sub-directories.

`landscapes/` contains data files and textures used for Stellarium's various landscapes. Each landscape has its own sub-directory. The name of this sub-directory is called the *landscape ID*, which is used to specify the default landscape in the main configuration file, or in script commands.

`skycultures/` contains constellations, common star names and constellation artwork for Stellarium's many sky cultures. Each culture has its own sub-directory in the skycultures directory.

`nebulae/` contains data and image files for nebula textures. In the future Stellarium may be able to support multiple sets of nebula images and switch between them at runtime. This feature is not implemented for version 0.15.0, although the directory structure is in place - each set of nebula textures has its own sub-directory in the nebulae directory.

---

[2]This is a very confusing default setting and in fact a security risk: Consider you receive an email with some file `funny.png.exe` attached. Your explorer displays this as `funny.png`. You double-click it, expecting to open some image browser with a funny image. However, you start some unknown program instead, and running this `.exe` executable program may turn out to be anything but funny!

[3]Windows Vista users who do not run Stellarium with administrator privileges should adjust the shortcut in the start menu to specify a different directory for screenshots as the Desktop directory is not writable for normal programs. Stellarium includes a GUI option to specify the screenshot directory.

[4]`http://www.mactipsandtricks.com/articles/Wiley_HT_appBundles.lasso`

stars/ contains Stellarium's star catalogues. In the future Stellarium may be able to support multiple star catalogues and switch between them at runtime. This feature is not implemented for version 0.15.0, although the directory structure is in place – each star catalogue has its own sub-directory in the stars directory.

data/ contains miscellaneous data files including fonts, solar system data, city locations, etc.

textures/ contains miscellaneous texture files, such as the graphics for the toolbar buttons, planet texture maps, etc.

ephem/ (optional) may contain data files for planetary ephemerides DE430 and DE431 (see 5.4.1).

If any file exists in both the installation directory and user directory, the version in the user directory will be used. Thus it is possible to override settings which are part of the main Stellarium installation by copying the relevant file to the user area and modifying it there.

It is recommended to add new landscapes or sky cultures by creating the relevant files and directories within the user directory, leaving the installation directory unchanged. In this manner different users on a multi-user system can customise Stellarium without affecting the other users.

5.3 The Main Configuration File

The main configuration file is read each time Stellarium starts, and settings such as the observer's location and display preferences are taken from it. Ideally this mechanism should be totally transparent to the user – anything that is configurable should be configured "in" the program GUI. However, at time of writing Stellarium isn't quite complete in this respect, despite improvements in each version. Some settings, esp. color values for lines, grids, etc. can only be changed by directly editing the configuration file.[5] This section describes some of the settings a user may wish to modify in this way, and how to do it.

If the configuration file does not exist in the *user directory* when Stellarium is started (e.g., the first time the user starts the program), one will be created with default values for all settings (refer to section 5 Files and Directories for the location of the user directory on your operating system). The name of the configuration file is config.ini[6].

The configuration file is a regular text file, so all you need to edit it is a text editor like **Notepad** on Windows, **Text Edit** on the Mac, or **nano/vi/gedit/emacs/leafpad** etc. on Linux.

A complete list of configuration file options and values may be found in appendix D.1 Configuration File.

5.4 Getting Extra Data

Stellarium is packaged with over 600 thousand stars in the normal program download, but much larger star catalogues may be downloaded in the *Tools* tab of the *Configuration* dialog.

5.4.1 Alternative Planet Ephemerides: DE430, DE431   0.15

By default, Stellarium uses the *VSOP87* planetary theory, an analytical solution which is able to deliver planetary positions for any input date. However, its use is recommended only for the year range $-4000\ldots+8000$. Outside this range, it seems to be usable for a few more millennia without too great errors, but with degrading accuracy.

Since V0.15 you can install extra data files which allow access to the numerical integration runs *DE430* and *DE431* from NASA's Jet Propulsion Laboratory (JPL). The data files have to be

---

[5]Color values can be edited interactively by the Text User Interface plugin (see 11.6).

[6]It is possible to specify a different name for the main configuration file using the --config-file command line option. See section 6 Command Line Options for details.

downloaded separately, and most users will likely not need them. DE430 provides highly accurate data for the years $+1550\ldots+2650$, while DE431 covers years $-13000\ldots+17000$, which allows e.g. archaeoastronomical research on Mesolithic landscapes. Outside these year ranges, positional computation falls back to VSOP87.

The integration of this feature is still experimental. As of V0.15, solar eclipses in antiquity seem to be slightly off. Please use JPL Horizon for quotable results.

To enable use of these data, download the files from JPL[7]:

Ephemeris	Filename	MD5 hash
DE430	linux_p1550p2650.430	707c4262533d52d59abaaaa5e69c5738
DE431	lnxm13000p17000.431	fad0f432ae18c330f9e14915fbf8960a

The files can be placed in a folder named ephem inside either the *installation directory* or the *user directory* (see 5.2). Alternatively, if you have them already stored elsewhere, you may add the path to config.ini like:

```
[astro]
de430_path = C:/Astrodata/JPL_DE43x/linux_p1550p2650.430
de431_path = C:/Astrodata/JPL_DE43x/lnxm13000p17000.431
```

For fast access avoid storing them on a network drive or USB pendrive!

You activate use of either ephemeris in the configuration panel (F2). If you activate both, preference will be given for DE430 if the simulation time allows it. Outside of the valid times, VSOP87 will always be used.

Acknowledgement

The optional use of DE430/431 has been supported by the ESA Summer of Code 2015 initiative.

---

[7]ftp://ssd.jpl.nasa.gov/pub/eph/planets/Linux/. (Also download from this directory if you are not running Linux!)

6. Command Line Options

Stellarium's behaviour can be modified by providing parameters to the program when it is called via the command line. See table for a full list:

Option	Option Parameter	Description
--help or -h	[none]	Print a quick command line help message, and exit.
--version or -v	[none]	Print the program name and version information, and exit.
--config-file or -c	config file name	Specify the configuration file name. The default value is config.ini. The parameter can be a full path (which will be used verbatim) or a partial path. Partial paths will be searched for inside the regular search paths unless they start with a ".", which may be used to explicitly specify a file in the current directory or similar. For example, using the option -c my_config.ini would resolve to the file <user directory>/my_config.ini whereas -c ./my_config.ini can be used to explicitly say the file my_config.ini in the current working directory.
--restore-defaults	[none]	Stellarium will start with the default configuration. Note: The old configuration file will be overwritten.
--user-dir	path	Specify the user data directory.

--screenshot-dir	path	Specify the directory to which screenshots will be saved.
--full-screen	yes or no	Over-rides the full screen setting in the config file.
--home-planet	planet	Specify observer planet (English name).
--altitude	altitude	Specify observer altitude in meters.
--longitude	longitude	Specify latitude, e.g. +53d58'16.65"
--latitude	latitude	Specify longitude, e.g. -1d4'27.48"
--list-landscapes	[none]	Print a list of available landscape IDs and exit.
--landscape	landscape ID	Start using landscape whose ID matches the passed parameter (dir name of landscape).
--sky-date	date	The initial date in `yyyymmdd` format.
--sky-time	time	The initial time in `hh:mm:ss` format.
--startup-script	script name	The name of a script to run after the program has started.
--fov	angle	The initial field of view in degrees.
--projection-type	ptype	The initial projection type (e.g. `perspective`).
--dump-opengl-details or -d	[none]	Dump information about OpenGL support to logfile. Use this is you have graphics problems and want to send a bug report.
--angle-mode or -a	[none]	Use ANGLE as OpenGL ES2 rendering engine (autodetect Direct3D version).[1]
--angle-d3d9 or -9	[none]	Force use Direct3D 9 for ANGLE OpenGL ES2 rendering engine.[1]
--angle-d3d11	[none]	Force use Direct3D 11 for ANGLE OpenGL ES2 rendering engine.[1]
--angle-warp	[none]	Force use the Direct3D 11 software rasterizer for ANGLE OpenGL ES2 rendering engine.[1]
--mesa-mode or -m	[none]	Use MESA as software OpenGL rendering engine.[1]
--safe-mode or -s	[none]	Synonymous to --mesa-mode.[1]
--fix-text or -t	[none]	Alternative way of creating the Info text, required on some systems.[2]

0.15   If you want to avoid adding the same switch every time when you start Stellarium from the command line, you can also set an environment variable STEL_OPTS with your default options.

---

[1] On Windows only

[2] E.g., Raspberry Pi 2 with Raspbian Jessie and VC4 drivers from February 2016. A bugfix should be available later in 2016.

6.1 Examples

- To start Stellarium using the configuration file, `configuration_one.ini` situated in the user directory (use either of these):

```
stellarium --config-file=configuration_one.ini
stellarium -c configuration_one.ini
```

- To list the available landscapes, and then start using the landscape with the ID "ocean"

```
stellarium --list-landscapes
stellarium --landscape=ocean
```

Note that console output (like --list-landscapes) on Windows is not possible.

7. Landscapes

GEORG ZOTTI

Landscapes are one of the key features that make Stellarium popular. Originally just used for decoration, since version 10.6 they can be configured accurately for research and demonstration in "skyscape astronomy", a term which describes the connection of landscape and the sky above [11]. Configured properly, they can act as reliable proxies of the real landscapes, so that you can take e.g. measurements of sunrise or stellar alignments [75], or prepare your next moonrise photograph, as though you were on-site.

In this chapter you can find relevant information required to accurately configure Stellarium landscapes, using panoramas created from photographs taken on-site, optionally supported by horizon measurements with a theodolite.

Creating an accurate panorama requires some experience with photography and image processing. However, great open-source tools have been developed to help you on the job. If you already know other tools, you should be able to easily transfer the presented concepts to those other tools.

7.1 Stellarium Landscapes

As of version 0.15, the available landscape types are:

polygonal A point list of measured azimuth/altitude pairs, used to define a sharp horizon polygon. The area below the horizon line is colored in a single color (Section 7.1.2).

spherical The simple form to configure a photo-based panorama: A single image is used as texture map for the horizon (Section 7.1.3).

old_style The original photo panorama. This is the most difficult to configure, but allows highest resolution by using several texture maps (Section 7.1.4).

fisheye Another 1-texture approach, utilizing an image made with a fisheye lens. This landscape suffers from calibration uncertainties and can only be recommended for decoration (Section 7.1.5).

A landscape consists of a `landscape.ini` plus the data files that are referenced from there, like a coordinate list or the textures. Those reside in a subdirectory of the `landscape` folder inside the Stellarium program directory, or, for own work, in a subdirectory of the `landscape` folder inside your Stellarium user data directory (see section 5.1).

Let us ssume we want to create a landscape for a place called Rosenburg. The location for the files of our new custom landscape *Rosenburg* depends on the operating system (see 5.1). Create a new subdirectory, and for maximum compatibility, use small letters and no spaces:

Windows `C:/Users/YOU/AppData/Roaming/Stellarium/landscapes/rosenburg`

Linux `~/.stellarium/landscapes/rosenburg`

Mac `$HOME/Library/Preferences/Stellarium/landscapes/rosenburg`

7.1.1 Location information

This optional section in `landscape.ini` allows automatic loading of site coordinates if this option is activated in the program GUI (see 4.4.4). For our purposes we should consider especially the coordinates in the location section mandatory!

```
[location]
planet = Earth
country = Austria
name = KGA Rosenburg
latitude = +48d38'3.3"
longitude = +15d38'2.8"
altitude = 266
light_pollution = 1
atmospheric_extinction_coefficient = 0.2
display_fog = 0
atmospheric_temperature = 10.0
atmospheric_pressure = 1013.0
```

Where:

planet Is the English name of the solar system body for the landscape.

latitude Is the latitude of site of the landscape in degrees, minutes and seconds. Positive values represent North of the equator, negative values South of the equator.

longitude Is the longitude of site of the landscape. Positive values represent East of the Greenwich Meridian on Earth (or equivalent on other bodies), Negative values represent Western longitude.

altitude Is the altitude of the site of the landscape in meters.

country (optional) Name of the country the location is in.

state (optional) Name of the state the location is in.

name (optional) Name of the location. This may contain spaces, but keep it short to have it fully visible in the selection box.

Since V0.11, there are a few more optional parameters that can be loaded if the according switch is active in the landscape selection panel. If they are missing, the parameters do not change to defaults.

light_pollution (optional) Light pollution of the site, given on the Bortle Scale (1: none ... 9: metropolitan; see Appendix B). If negative or absent, no change will be made.

atmospheric_extinction_coefficient (optional, no change if absent.) Extinction coefficient (mag/airmass) for this site.

atmospheric_temperature (optional, no change if absent.) Surface air temperature (Degrees Celsius). Used for refraction. Set to -1000 to explicitly declare "no change".

atmospheric_pressure (optional, no change if absent.) Surface air pressure (mbar; would be 1013 for "normal" sea-level conditions). Used for refraction. Set to -2 to declare "no change", or -1 to compute from altitude.

display_fog (optional, -1/0/1, default=-1) You may want to preconfigure setting 0 for a landscape on the Moon. Set -1 to declare "no change".

7.1.2 Polygonal landscape

This landscape type has been added only recently (since 0.13) to allow the use of measured horizons. Users of **Cartes du Ciel**[1] will be happy to hear that the format of the list of measurements is compatible.

This is the technically simplest of the landscapes, but may be used to describe accurately measured horizon lines. The file that encodes horizon altitudes can also be used in all other landscape types. If present there, it will be used to define object visibility (instead of the opacity of the landscape photo textures) and, if horizon_line_color is defined, will be plotted.

There is a small caveat: Sometimes, there may appear vertical lines from some corners towards the zenith or the mathematical horizon, e.g. if there is a vertex including azimuth 0 or 180. If this irritates you, just offset this azimuth minimally (e.g., 180.00001).

The landscape.ini file for a polygonal type landscape looks like this (this example is based on the *Geneve* landscape which was borrowed from Cartes du Ciel and comes with Stellarium):

```
[landscape]
name = Geneve
type = polygonal
author = Georg Zotti; Horizon definition by Patrick Chevalley
description = Horizon line of Geneve.
              Demonstrates compatibility with
              horizon descriptions from Cartes du Ciel.
polygonal_horizon_list = horizon_Geneve.txt
polygonal_angle_rotatez = 0
ground_color = .15,.45,.45
horizon_line_color = .75,.45,.45
```

Where:

name appears in the landscape tab of the configuration window.

type identifies the method used for this landscape. polygonal in this case.

author lists the author(s) responsible for images and composition.

description gives a short description visible in the selection panel. The text can be superseded by optional description.<lang>.utf8 files.

polygonal_horizon_list is the name of the horizon data file for this landscape.

polygonal_horizon_list_mode (optional) the two first columns in the list are numbers: azimuth and altitude or zenith distance, in either degrees or radians or gradians(gon). The value must be one of azDeg_altDeg, azDeg_zdDeg, azRad_altRad, azRad_zdRad, azGrad_altGrad, azGrad_zdGrad. Default: azDeg_altDeg

polygonal_angle_rotatez (optional, default=0) Angle (degrees) to adjust azimuth. This may be used to apply a (usually) small offset rotation, e.g. when you have measured the horizon in a grid-based coordinate system like UTM and have to compensate for the meridian convergence.

---

[1]SkyChart / Cartes du Ciel planetarium: http://www.ap-i.net/skychart/en/start

ground_color (optional, default=0,0,0, i.e., black) Color for the area below the horizon line. Each R,G,B component is a float within 0..1.

horizon_line_color (optional, default: invisible) used to draw a polygonal horizon line. Each R,G,B component is a float within 0..1.

minimal_brightness (optional) Some minimum brightness to keep landscape visible. Default=-1, i.e., use minimal_brightness from the [landscape] section in the global config.ini.

minimal_altitude (optional, default=-2) Some sky elements, e.g. stars, are not drawn below this altitude to increase performance. Under certain circumstances you may want to specify something else here. (since V0.14)

polygonal_horizon_inverted (optional, default=false) In rare cases like horizon lines for high mountain peaks with many negative horizon values this should be set to true. (since V0.15)

7.1.3   Spherical landscape

This method uses a more usual type of panorama – the kind which is produced directly from software such as **autostitch** or **Hugin**[2]. The *Moon* landscape which comes with Stellarium provides a minimal example of a landscape.ini file for a spherical type landscape:

```
[landscape]
name = Moon
type = spherical
maptex = apollo17.png
```

A more elaborate example is found with the *Grossmugl* landscape:

```
[landscape]
name = Grossmugl
type = spherical
author = Guenther Wuchterl, Kuffner-Sternwarte.at;
        Lightscape: Georg Zotti
description = Field near Leeberg, Grossmugl (Riesentumulus),
             Austria - Primary Observing Spot of the Grossmugl
             Starlight Oasis - http://starlightoasis.org
maptex = grossmugl_leeberg_crop11.25.png
maptex_top=11.25
maptex_fog = grossmugl_leeberg_fog_crop22.5.png
maptex_fog_top = 22.5
maptex_fog_bottom = -22.5
maptex_illum = grossmugl_leeberg_illum_crop0.png
maptex_illum_bottom = 0
angle_rotatez=-89.1
minimal_brightness = 0.0075
polygonal_horizon_list = horizon_grossmugl.txt
polygonal_angle_rotatez=0
horizon_line_color =  .75,.45,.45
minimal_altitude = -1
```

Where:

name appears in the landscape tab of the configuration window. This name may be translated.

type identifies the method used for this landscape. spherical in this case.

-----

[2]http://hugin.sourceforge.net/

author lists the author(s) responsible for images and composition.

description gives a short description visible in the selection panel. The text will be superseded by optional description.<lang>.utf8 files.

maptex is the name of the image file for this landscape.

maptex_top (optional; default=90) is the altitude angle of the top edge.

maptex_bottom (optional; default=-90) is the altitude angle of the bottom edge. Usually you will not require this, or else there will be a hole at your feet. ;-)

maptex_fog (optional; default: no fog) is the name of the fog image file for this landscape.

maptex_fog_top (optional; default=90) is the altitude angle of the top edge of the fog texture. Useful to crop away parts of the image to conserve texture memory.

maptex_fog_bottom (optional; default=-90) is the altitude angle of the bottom edge.

maptex_illum (optional; default: no illumination layer) is the name of the nocturnal illumination/light pollution image file for this landscape.

maptex_illum_top (optional; default=90) is the altitude angle of the top edge, if you have light pollution only close to the horizon.

maptex_illum_bottom (optional; default=-90) is the altitude angle of the bottom edge.

angle_rotatez (optional, default=0) Angle (degrees) to adjust azimuth. If 0, the left/right edge is due east.

tesselate_rows (optional, default=20) This is the number of rows for the maptex. If straight vertical edges in your landscape appear broken, try increasing this value, but higher values require more computing power. Fog and illumination textures will have a similar vertical resolution.

tesselate_cols (optional, default=40) If straight horizontal edges in your landscape appear broken, try increasing.

polygonal_horizon_list (optional) is the name of the (measured) horizon data file for this landscape. Can be used to define the exact position of the horizon. If missing, the texture can be queried for horizon transparency (for accurate object rising/setting times)

polygonal_horizon_list_mode (optional) see 7.1.2

polygonal_angle_rotatez (optional, default=0) see 7.1.2

horizon_line_color see 7.1.2

minimal_brightness see 7.1.2

minimal_altitude (optional, default=-2) Some sky elements, e.g. stars, are not drawn below this altitude for efficiency. Under certain circumstances (e.g. for space station panoramas where you may have sky below your feet, or for deep valleys/high mountains, you may want to specify something else here. (since V0.14)

To save texture memory, you can trim away the transparent sky and define the angle maptex_top. Likewise, fogtex_top, fogtex_bottom, maptex_illum_top and maptex_illum_top. You should then stretch the texture to a full power of 2, like 4096×1024 (but note that some hardware is even limited to 2048 pixels). The easiest method to create perfectly aligned fog and illumination layers is with an image editor that supports layers like the **GIMP** or **Photoshop**. Fog and Light images should have black background.

7.1.4 High resolution ("Old Style") landscape

The old_style or multiple image method works by having the 360°panorama of the horizon (without wasting too much texture memory with the sky) split into a number of reasonably small *side textures*, and a separate *ground texture*. This has the advantage over the single-image method that the detail level of the horizon can be increased without ending up with a single very large image file, so this is usable for either very high-resolution panoramas or for older hardware with limited capabilities. The ground texture can be a different resolution than the side textures. Memory usage

Figure 7.1: Old_style landscape: eight parts delivering a high-resolution panorama. The bottom (ground) texture, drawn on a flat plane, is not shown here.

may be more efficient because there are no unused texture parts like the corners of the texture file in the fish-eye method. It is even possible to repeat the horizon several times (for purely decorative purpose). The side textures are mapped onto curved (spherical ring or cylinder) walls (Fig. 7.1).

On the negative side, it is more difficult to create this type of landscape – merging the ground texture with the side textures can prove tricky. (**Hugin** can be used to create also this file, though. And on the other hand, you can replace this by something else like a site map.) The contents of the landscape.ini file for this landscape type is also somewhat more complicated than for other landscape types. Here is the landscape.ini file which describes our *Rosenburg* landscape[3]:

```
[landscape]
name = KGA Rosenburg
author = Georg Zotti, VIAS/ASTROSIM
description = KGA Rosenburg
type = old_style
nbsidetex = 8
tex0 = Horiz-0.png
tex1 = Horiz-1.png
tex2 = Horiz-2.png
tex3 = Horiz-3.png
tex4 = Horiz-4.png
tex5 = Horiz-5.png
tex6 = Horiz-6.png
tex7 = Horiz-7.png
nbside = 8
side0 = tex0:0:0:1:1
side1 = tex1:0:0:1:1
side2 = tex2:0:0:1:1
```

---

[3]the groundtex grassground.png mentioned here has been taken from the *Guereins* landscape.

```
side3 = tex3:0:0:1:1
side4 = tex4:0:0:1:1
side5 = tex5:0:0:1:1
side6 = tex6:0:0:1:1
side7 = tex7:0:0:1:1
groundtex = grassground.png
ground = groundtex:0:0:1:1
nb_decor_repeat = 1
decor_alt_angle =   82
decor_angle_shift = -62
; Rotatez deviates from -90 by the Meridian Convergence.
; The original landscape pano is grid-aligned, not north-aligned!
decor_angle_rotatez =   -90.525837223
ground_angle_shift = -62
ground_angle_rotatez =   44.474162777
draw_ground_first = 1
fogtex = fog.png
fog_alt_angle = 20
fog_angle_shift = -3
fog = fogtex:0:0:1:1
calibrated = true
[location]
planet = Earth
latitude = +48d38'3.3"
longitude = +15d38'2.8"
altitude = 266
light_pollution = 1
atmospheric_extinction_coefficient = 0.2
display_fog = 0
atmospheric_temperature = 10.0
atmospheric_pressure = 1013.0
```

Where:

name is the name that will appear in the landscape tab of the configuration window for this
landscape

type should be old_style for the multiple image method.

author lists the author(s) responsible for images and composition.

description gives a short description visible in the selection panel. The text will be superseded
by optional description.<lang>.utf8 files.

nbsidetex is the number of side textures for the landscape.

tex0 ... tex<nbsidetex-1> are the side texture file names. These should exist in the
textures / landscapes / landscape directory in PNG format.

light0 ... light<nbsidetex-1> are optional textures. If they exist, they are used as overlays
on top of the respective tex<...> files and represent nocturnal illumination, e.g. street
lamps, lit windows, red dots on towers, sky glow by city light pollution, ... Empty (black)
panels can be omitted. They are rendered exactly over the tex<...> files even when the
PNG files have different size. If you need your light pollution higher in the sky, you must
use a spherical or fisheye landscape.

nbside is the number of side textures

side0 ...side<nbside-1> are the descriptions of how the side textures should be arranged in the program. Each description contains five fields separated by colon characters (:). The first field is the ID of the texture (e.g. tex0), the remaining fields are the texture coordinates (x0:y0:x1:y1) used to place the texture in the scene. If you want to use all of the image, this will just be 0:0:1:1.

groundtex is the name of the ground texture file. (This could also be a diagram e.g. indicating the mountain peaks!)

fogtex is the name of the texture file for fog in this landscape. Fog is mapped onto a simple cylinder.[4] Note that for this landscape, accurate overlay of fog and landscape is only guaranteed if calibrated=true and tan_mode=true.

nb_decor_repeat is the number of times to repeat the side textures in the 360 panorama. (Useful photo panoramas should have 1 here)

decor_alt_angle (degrees) is the vertical angular extent of the textures (i.e. how many degrees of the full altitude range they span).

decor_angle_shift (degrees) vertical angular offset of the scenery textures, at which height the bottom line of the side textures is placed.

decor_angle_rotatez (degrees) angular rotation of the panorama around the vertical axis. This is handy for rotating the landscape so North is in the correct direction. Note that for historical reasons, a landscape with this value set to zero degrees has its leftmost edge pointing towards east.

ground_angle_shift (degrees) vertical angular offset of the ground texture, at which height the ground texture is placed.

ground_angle_rotatez (degrees) angular rotation of the ground texture around the vertical axis. When the sides are rotated, the ground texture may need to be rotated as well to match up with the sides.

fog_alt_angle (degrees) vertical angular size of the fog cylinder - how fog looks. Accurate vertical size requires calibrated=true.

fog_angle_shift (degrees) vertical angular offset of the fog texture - at what height is it drawn. Accurate vertical placement requires calibrated=true.

draw_ground_first if 1 the ground is drawn in front of the scenery, i.e. the side textures will overlap over the ground texture.

calibrated (optional). New since V0.10.6: Only if true, decor_alt_angle etc. really work as documented above. The (buggy) old code was left to work with the landscapes already existing. Note that with "uncalibrated" landscapes, sunrise computations and similar functionality which requires an accurate horizon line will not work.

tan_mode (optional, not used in this file). If true, the panorama image must be in in cylindrical, not equirectangular projection. Finding decor_alt_angle and decor_angle_shift may be a bit more difficult with this, but now (V0.13) works also with calibrated. A fog image created as overlay on the pano will be perfectly placed.

decor_angle_rotatez angular rotation of the scenery around the vertical axis. This is handy for rotating the landscape so North is in the correct direction. If 0, the left edge of tex0 is due east.

ground_angle_shift vertical angular offset of the ground texture, at which height the ground texture is placed. Values above -10 are not recommended for non-photographic content (e.g., a map) due to high distortion.

ground_angle_rotatez angular rotation of the ground texture around the vertical axis. When the sides are rotated, the ground texture may need to be rotated as well to match up with the sides. If 0, east is up. if North is up in your image, set this to 90. Note that adjustments of

---

[4]In very wide-angle views, the fog cylinder may become visible in the corners.

> `decor_angle_rotatez` require adjustments of this angle in the opposite direction!

`fog_alt_angle` vertical angular size of the fog cylinder.

`fog_angle_shift` vertical angular offset of the fog cylinder.

`draw_ground_first` if 1, the ground is drawn before the sides, i.e. the side textures may overlap
the ground texture if `ground_angle_shift` > `decor_angle_shift`.

`polygonal_horizon_list` (optional) see 7.1.2

`polygonal_horizon_list_mode` (optional) see 7.1.2

`polygonal_angle_rotatez` (optional, default=0) see 7.1.2

`horizon_line_color` see 7.1.2

`minimal_brightness` see 7.1.2

`minimal_altitude` see 7.1.2

7.1.5　Fisheye landscape

The *Trees* landscape that is provided with Stellarium is an example of the single fish-eye method,
and provides a good illustration. The centre of the image is the spot directly above the observer
(the zenith). The point below the observer (the nadir) becomes a circle that just touches the edges
of the image. The remaining areas of the image (the corners outside the circle) are not used.

The image file (Fig. 7.2) should be saved in PNG format with alpha transparency. Whereever
the image is transparent Stellarium will render the sky.

The `landscape.ini` file for a fish-eye type landscape looks like this (this example is based on
the *Trees* landscape which comes with Stellarium):

```
[landscape]
name = Trees
type = fisheye
author = Robert Spearman. Light pollution image: Georg Zotti
description = Trees in Greenlake Park, Seattle
maptex = trees_512.png
maptex_illum = trees_illum_512.png
maptex_fog = trees_fog_512.png
texturefov = 210
angle_rotatez = 17
tesselate_rows = 28
tesselate_cols = 60
```

Where:

`name` appears in the landscape tab of the configuration window.

`type` identifies the method used for this landscape. `fisheye` in this case.

`author` lists the author(s) responsible for images and composition.

`description` gives a short description visible in the selection panel. The text will be superseded
by optional `description.<lang>.utf8` files.

`maptex` is the name of the image file for this landscape.

`maptex_fog` (optional) is the name of the fog image file for this landscape.

`maptex_illum` (optional) is the name of the nocturnal illumination/light pollution image file for
this landscape.

`texturefov` is the field of view that the image covers in degrees.

`angle_rotatez` (optional) Angle (degrees) to adjust azimuth.

`tesselate_rows` (optional, default=20) If straight edges in your landscape appear broken, try
increasing.

Figure 7.2: Texture for the *Trees* Fisheye landscape.

`tesselate_cols` (optional, default=40) If straight edges in your landscape appear broken, try increasing.

`polygonal_horizon_list` (optional) see 7.1.2

`polygonal_horizon_list_mode` (optional) see 7.1.2

`polygonal_angle_rotatez` (optional, default=0) see 7.1.2

`horizon_line_color` see 7.1.2

`minimal_brightness` see 7.1.2

`minimal_altitude` see 7.1.2

7.1.6 Description

The short `description` entry in `landscape.ini` will be replaced by the contents of an optional file `description.<LANG>.utf8`. `<LANG>` is the ISO 639-1 language code, or its extension which contains language and country code, like `pt_BR` for Brazilian Portuguese. The long description requires the file `description.en.utf8`, this is en=english text with optional HTML tags for sections, tables, etc. You can also have embedded images in the HTML (Views of sacred landscapes, other informative images, . . . ?), just make them PNG format please. The length of the description texts is not limited, you have room for a good description, links to external resources, whatever seems suitable.

If you can provide other languages supported by Stellarium, you can provide translations yourself, else Stellarium translators *may* translate the English version for you. (It may take years though.) The file ending .utf8 indicates that for special characters like ÄÖÜßáé you should use UTF8 encoding. If you write only English/ASCII, this may not be relevant.

7.1.7 Gazetteer   0.14

An optional feature for landscapes is a gazetteer function, i.e., labels for landscape features. The *Grossmugl* landscape demonstrates an example and should be self-explanatory. This is again multilingual, so the files are called gazetteer.<LANG>.utf8.

```
# demo gazetteer for Grossmugl landscape.
# Can be used to better describe the landscape,
# i.e. show labels on landscape features.
# Fields must be separated by vertical line,
# label must not have such a vertical line.
# Comments have this hash mark in first column.
# coordinates in degrees from true North.
# line towards zenith draws a single line strictly upward.
# label is centered on line endpoint.
# Azimuth | Altitude | degrees        | azimuth | label
#         |          | towards zenith | shift   |
113.66   | 5.5      | 4              | -6      | Leeberg
35       | 1.5      | 2.5            | 0       | Grossmugl
335      | 2        | 2              | 0       | Steinabrunn
305      | 2        | 1              | 0       | Ringendorf
180      | 2        | 2              | 0       | Vienna (30km)
135      | 2        | 0.5            | 0       | Wind power plant Strasshof
```

Figure 7.3: Zenit "Horizon 202" panorama camera with rotating lens for 35mm film. (Source: Wikipedia, "Horizon202" by BillC - Own Work. Licensed under CC BY-SA 3.0 via Wikimedia Commons - `https://commons.wikimedia.org/wiki/File:Horizon202.jpg#mediaviewer/File:Horizon202.jpg`)

7.2 Creating Panorama Photographs for Stellarium

7.2.1 Panorama Photography

Traditional film-based panorama photography required dedicated cameras with curved film holders and specialized lenses (Figure 7.3).

Digital photography has brought a revolution also in this field, and it has become quite easy to create panoramas simply by taking a series of photographs with a regular camera on the same spot and combining them with dedicated software.

A complete panorama photo visually encloses the observer like the mental image that astronomers have been using for millennia: the celestial sphere. If we want to document the view, say, in a big hall like a church, optimal results will be gained with a camera on a tripod with a specialized panorama head (Figure 7.4) which assures the camera rotates around the *entrance pupil*[5] of the lens in order to avoid errors by the parallax shift observed on photographs taken on adjacent but separate positions.

Often however, both the upper half of the observer's environment (the sky) and the ground the photographer is standing on, are regarded of lesser importance, and only a series of laterally adjacent photographs is taken and combined into a cylindrical or spherical ring that shows the landscape horizon, i.e., where ground and sky meet. If the closest object of interest is farther away that a few metres, requirements on parallax avoidance are far less critical, and the author has taken lots of landscape panoramas with a camera on the usual tripod screw, and even more entirely without a tripod. However, any visible errors that are caused by a shifted camera will require more effort in postprocessing.

When you have no tripod, note that *you must not rotate the camera on your outstretched arm!* Rather, the camera's entrance pupil must be rotated, so you should appear to dance around the

---

[5]In many references you will find "Nodal Point" mentioned here. But see these: `http://en.wikipedia.org/wiki/Cardinal_point_%28optics%29#Nodal_points`, `http://web.archive.org/web/20060513074042/http://doug.kerr.home.att.net/pumpkin/Pivot_Point.pdf`, `http://www.janrik.net/PanoPostings/NoParallaxPoint/TheoryOfTheNoParallaxPoint.pdf`

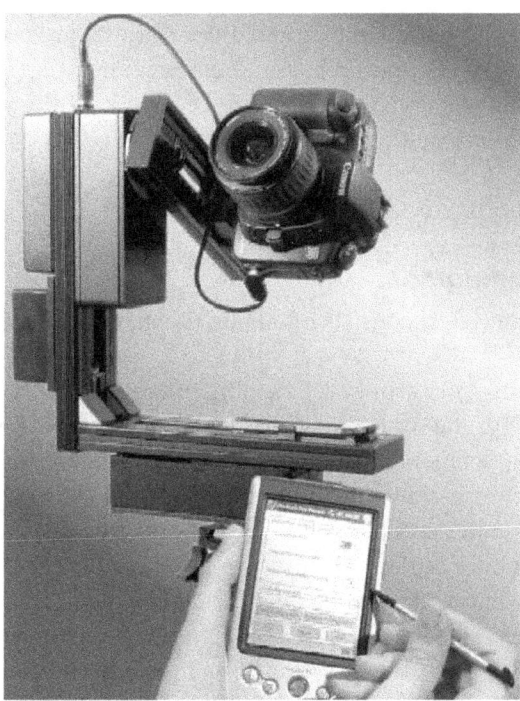

Figure 7.4: Automated panorama head. (Source: Wikipedia `https://de.wikipedia.org/wiki/Panoramakamera#mediaviewer/File:Rodeon_vr_head_01.jpg`)

camera!

The images should match in brightness and white balance. If you can shoot in RAW, do so to be able to change white balance later. If the camera can only create JPG, ensure you have set the camera to a suitable white balance before taking the photos and not to "auto", because this may find different settings and thus give colour mismatches. Exposure brightness differences can be largely removed during stitching, but good, well-exposed original shots always give better results.

As a general recommendation, the images of a panorama should be taken from left to right, else please accordingly invert some of the instructions given below.

There are several panorama making programs. Often they are included in the software that comes with a digital camera and allow the creation of simple panoramas. Other software titles are available for purchase. However, there is one cost-free open-source program that does everything we need for our task, and much more:

7.2.2 Hugin Panorama Software

Hugin[6], named after one of the ravens that sits on Odin's shoulder and tells him about the world, is a user-friendly catch-all package with graphical user interface that allows creating panoramas with a single application. Actually, **Hugin** is a GUI application which calls several specialized sub-programs with fitting parameters. The instructions are based on **Hugin V2014.0** and **2015.0**.

Typically digital images come in JPG format with information about camera, lens, and settings stored in invisible metadata in the EXIF format. When **Hugin** reads such images, it can automatically derive focal length, field of view, and exposure differences (exposure time, aperture, color balance) to create panoramas as easily as possible.

After starting **Hugin** for the first time, select Interface ⟩ Expert to release several options not visible to "beginners". In the Preferences dialog (Files ⟩ Preferences), edit number of CPU to

---

[6]`http://hugin.sourceforge.net/`

match the number of cores in your computer and allow parallel processing. E.g., if you have an Intel Core-i7, you usually can set up to 8 cores (4 cores with hyperthreading; but maybe leave one core for your other tasks while you wait for a processing job?). If your PC is equipped with a modern programmable graphics card, you can enable its use in the | Programs | tab with activating "Use GPU for remapping".

After that, we are ready for creating our panoramas.

7.2.3 Regular creation of panoramas

The graphical user interface (GUI) consists of a main menu, symbols, and 4 tabs. We start on the tab Photos.

- | Add images... | Opens a file browser. Select the images which you want to stitch. Usually, lens data (focal length, field of view, ...) are read from the EXIF data. If those are not available (e.g. cheap cameras, images scanned from film), you can enter those data on loading or later. The images are now listed in the file list, and you can edit image parameters by marking one or more, and then choosing from the context menu which you get from pressing the right mouse button. In case you have used different lenses (or inadvertently used different focal lengths of a zoom lens), you can assign separate lenses to the images. Caveat: If you have resized the images, or produced copied on your RAW converter with non-native resolution, the Field of View (FoV) in **Hugin** may be misidentified. You must edit lens parameters and fill in the field of view from a full-size image. Else the first round of optimisation will run into unsolvable trouble.
- Select one image as *position anchor* (usually the center image), and one as *exposure anchor* (this can be the same image). For our purpose, *the anchor image should face south.*
- Next, we must find common feature points. The next field below provides the required settings. It is recommended to use the **CPFind** command. To avoid finding control points in (moving) clouds, select setting | Hugin's CPFind + Celeste |[7]. Then press | Create control points |. This opens a dialog box in which you can see output of the selected feature point extractor. It should finish with a box telling you the number of identified points. In rare cases some images cannot be linked to others, you will have to manually add or edit feature points in those cases.
- Now it's time to start optimisations. On the | Geometric Optimimisation | combo, start with the button | Positions, incremental from anchor |, and press | Calculate |. Moments later, a first rough match is available for inspection.
- First open the Preview window (press | Ctrl |+| P | or click the blue icon). Assumed your images cover the full horizon, the window shows an equirectangular area (360 degrees along the horizon and 180 degrees from zenith to nadir). The anchor image should be close to the image center, and the other images should be already well-aligned to both sides. You can set the exact center point by clicking it in the image. If the horizon appears badly warped, use the right mouse key and click on the horizon roughly near −90 or +90 degrees (halfway to the left or right).
- Open the OpenGL preview window (press | Ctrl |+| Shift |+| P | or click the blue icon with GL inside). This panel provides several important views:
  - The | Preview | tab is similar to the non-OpenGL preview. You can display an overlay of the control points, which are colored according to match quality. Also, with button | Identify | activated, you see the overlapping image frames when you move the mouse over the image.
  - The | Layout | tab helps finding links between images.
  - The | Move/Drag | dialog may help to interactively adjust a panorama.

---

[7]If you forget this, you can remove cloud points by calling **Celeste** in the control point editor later

Sometimes the preview image may however be distorted and unusable.

- Open the *Control Points Table* dialog (press F3 or click the "table" button). Here you see the points listed which link two images. Clicking a column label sorts by this column. It is recommended that only neighboring overlapping images should be included here. If you have very large overlap, it is possible that points are found between two images which are not directly adjacent. In the OpenGL preview window, you can use the Preview or the Layout tabs to identify those image pairs. Such points should be deleted. In the point table, click on columns "Right Img.", then "Left Img.", and then find pairs like 0/2, 1/3, 2/4 etc. Mark those lines, and delete the points.
- To re-run the optimisation, press the double-arrow icon or the calculate button in the Optimise/Geometric area.

Preliminary Geometric Optimisation

Now the (usually) longest part begins: Iterative optimisation of the photo matchpoints. If your images were taken on a panorama tripod head, there should only be very few bad matchpoints, e.g. those found on persons or clouds[8] which have moved between photos. For handheld photos, the following considerations should be observed.

The most important line which we want to create in all perfection is the visible horizon, where sky and earth meet. The foreground, usually grassy or rocky, is of lesser interest, and stitching errors in those areas may not even be relevant.

Therefore, matchpoints with large errors in the foreground can be safely removed, while, if necessary, points on the horizon should be added manually. Use the Control Points tab, select adjacent images (start with 0 on the left and 1 on the right side), and delete the worst-fitting matchpoints closest to the camera (near the bottom of the images). We now start a long phase of re-optimizing and deletion of ill-matching points as long as those are far from the horizon. When all near matchpoints are deleted, the result should already look not too bad.

For continued optimisation, the number of parameters to optimize can be extended. To begin, I recommend Positions and View (y, p, r, v), which may find a new focal length slightly different from the data in the EXIF tags. Again, delete further foreground points. If after a few rounds you still have bad point distances, try Positions and Barrel Distortion (y, p, r, b) to balance distortion by bad optics, or even go up to Everything without translation. Optimisation can only reach perfect results if you did not move between exposures. Else, find a solution which shows the least error.

In case you took your photos not on a tripod and moved too much, you may even want to play with the translation options, but errors will be increasingly hard to avoid.

Using Straight Edges as Guides

If the panorama contains straight lines like vertical edges of buildings, these can be used to automatically get a correctly levelled horizon: Vertical lines are mapped to vertical lines in equirectangular panos! In the Control Points tab, select the image with the vertical edge in both subframes, and mark points on the vertical edge. (switch off auto-estimate!). Likewise, horizontal lines may help, but make sure lines like rooves are perpendicular to your line of view, else the perspective effect causes an inclination.

Multi-ring Panoramas

If you are trying to create a panorama with several rings (horizon, one or two rings below, and nadir area), you must try to create/keep control points that best give a result without visible seams. In this case, and esp. if you have only used a regular tripod or even dared to go for a free-handed panorama, you may observe that it is best to remove control points in neighboring photos in the

---

[8]You should have created control points with the **Celeste** option!

lower rings, but keep only the "vertical" links between images with similar azimuth.

In total, and if the foreground is not important but only grassy or sandy, the rule of thumb is that the horizon images must be strongly linked with good quality (small errors), while images in the lower rings should be linked mostly to their respective upper photos, but not necessarily to the images to its sides. The resulting panorama will then show a good horizon line, while stitching artifacts in a grassy or otherwise only decorative ground will usually be acceptable and can, if needed, be camouflaged in postprocessing.

This optimisation and editing of control points is likely a longish iterative process, and these are the late night hours where you will finally wish you had used a panorama head...

Masking

If you have images with overlapping areas, you can usually not force **Hugin** to take pixels from the image which you find best. you can however mask off an area from an image which you don't want to see in the output under any circumstances, e.g. a person's arm or foot in one image. Just open the image in the Mask tab and either press Add new mask and draw the mask polygon covering the unwanted area, or use the crop settings to define rectangular areas to use.

Exposure disbalance

In the Photos tab, select Photometric parameters on the right side. The EV column lists the *Exposure Value*. If you see disbalance here and in the preview window, you can run a photometric optimisation with the lowest button on the Photos tab. Simply select Low dynamic range and press Calculate . The preview should now show a seamless image. If all else fails, you can edit the EV values directly.

Advanced photographers may want to correct exposures in their RAW images before creating JPG or TIF images to combine with **Hugin**. This unfortunately may create exposure disbalance because the EXIF tags may not be adjusted accordingly, so based on different exposure/f-stop conbinations **Hugin** may think it has to re-balance the values. In these cases, don't run the photometric optimiser. Some image exposure values have to be changed manually, and the effect supervised in the preview window. Usually the smooth blending in the subprogam **enblend** called by **Hugin** will hide remaining differences.

Stitching

When you are happy with the panorama in the preview window and the matchpoints promise a good fit, it is time to finally create the panorama image. **Hugin** can create a large number of different projections which all have their application. For Stellarium, we can only use the equirectangular projection. You still have 2 options:

spherical landscapes (see 7.1.3) require single equirectangular images, the maximum size depends on your graphics hardware and **Qt** limitations and is likely not larger than 8192×4096 pixels.

old_style landscapes (see 7.1.4) can use several textures for the ring along the horizon, and one image for the nadir zone. If you need high resolution, you should aim for creating this one.

Sometimes, creating the nadir zone is difficult: this is where usually the view is blocked by the tripod, and we are not interested in views of tripod or our own feet. For our purpose it is usually enough to fill in the feet area using the clone stamp, or a monochrome color, or, for old_style landscapes, you can instead insert an oriented site map or wind rose.

There is a button create optimal size in **Hugin**. It may recommend a panorama width around 13.000 pixels for an average camera and photos taken with a wide-angle lens. Increasing this size will most likely not lead to higher optical resolution! The panorama width which you can most usefully create depends on the resolution of the source images (which leads to the result given by **Hugin**) and on your needs. If you need arcminute resolution, you would aim for

$360 \times 60 = 21600$ pixels, which cannot be loaded into graphics memory in a single piece, i.e., is too large for Stellarium, and must be configured as old_style landscape. In this case, 10 or 11 tiles of 2048×2048 pixels (totalling 20480 or 22528 pixels) is the closest meaningful setting, i.e., you could create an image of 20480 pixels width and cut this into usable pieces. Usually, a size of 4096×2048 or 8192×4096 pixels (for better computers) is enough, and can be used in a spherical landscape.

We have to edit the file after stitching, therefore select creation of an image in the TIFF format. LZW compression is non-lossy, so use this to keep file size reasonably small.

For regular images, it is enough to create "Exposure corrected, low dynamic range". If you have a problem with persons that have moved between your images, you may want to post-process the final result with import of the distorted sub-images and manually defining the best blending line. For this, find the "Remapped Images" group and again activate "Exposure corrected, low dynamic range".

Now, press the $\boxed{\text{Stitch!}}$ button in the lower right corner. This opens a helper program which supervises the stitching process. Depending on your computer and size of the image, it will require a few minutes of processing.

In case stitching fails with a cryptic error message, try to add the option --fine-mask to the **enblend** options.

Store a copy of the **Hugin** project file to always be able to go back to the settings you used to create the last panorama. We will get back to it when we want to make a truly calibrated panorama (see 7.3.3).

7.3 Panorama Postprocessing

The image created has to be further processed to be used in Stellarium. The most obvious change is the need for a transparent sky, which we can easily create in programs like **Adobe Photoshop** or the free and open-source **GIMP**. I will describe only the free and open-source solution.

After that, we have to bring the image into shape for Stellarium, which may include some trimming. While we could also slice an image with interactive tools, higher accuracy and repeatable results can be achieved with command-line programs, which makes the **ImageMagick** suite the tool of our choice.

7.3.1 The GIMP

The **GIMP** (GNU Image Manipulation Program) has been developed as free alternative to the leading commercial product, **Adobe Photoshop**. While it may look a bit different, basic concepts are similar. Not everybody can (or wants to) afford **Photoshop**, therefore let's use the **GIMP**.

Like **Photoshop**, the **GIMP** is a layer-aware image editor. To understand the concept, it is easiest to imagine you operate on a growing stack of overhead slides. You can put a new transparent slide ("layer") on top of the stack and paint on this without modifying the lower layers.

A few important commands:

Zooming $\boxed{\text{Ctrl}}$ + $\boxed{\text{Mouse Wheel}}$

Layer visibility and transparency Make sure to have layer dialog shown ($\boxed{\text{Windows}} \gg \boxed{\text{Dockable Dialogs}}$).
> A gray bar indicates opacity for the currently active layer. Note the mouse cursor in this opacity bar (often also called transparency bar): near the top of the bar the upward pointer immediately sets percentage. A bit lower the pointer looks different and can be used for fine-tuning.

The most obvious postprocessing need for our panorama is making the sky transparent. The optimal tool usually is the "Fuzzy Select", which is equivalent to the "Magic Wand" tool in

Photoshop. Simply mark the sky, and then delete it. The checkerboard background indicates transparent pixels.

It sometimes helps to put an intensive bright red or blue background layer under the panorama photo to see the last remaining clouds and other specks. In the layer dialog, create a new layer, bucket-fill with blue or red, and drag it in the layer dialog below the pano layer. Write-protect this layer, work on the image layer, and before exporting the image layer with transparent sky to PNG, don't forget to switch off the background.

We need this layer functionality especially to align the panorama on a calibration grid, see section 7.3.3.

7.3.2   ImageMagick

ImageMagick (IM)[9] can be described as "Swiss Army Knife of image manipulation". It can do most operations usually applied to images in a GUI program, but is called from the command line. This allows also to include **IM** in your own command scripts[10]. We will use it to do our final cut and resize operations. I cannot give an exhaustive tutorial about more than a few of **IM**'s functions, but the commands given here should be enough for our purpose.

To open a command window (console, a.k.a. DOS window), press the Windows key and enter cmd, then press ⏎ . (On Linux and Mac, you surely know how to open a console window.)

There are some things you might need to know:

- The command line is not your enemy, but a way to call expert tools.
- The Windows command line processor **cmd.exe** is far from user friendly.
- There are remedies and alternatives. See notes on **clink** (7.4.3) for a considerable improvement, and **Cygwin** (7.4.4) for experts.

Command-line magick for spherical landscapes

Let's start with the commands for final dressing of an equirectangular panorama to be used as spherical landscape which has been created in size 4096×2048, but where you have seen that nothing interesting is in the image above 11.25°. This means we can cut away the sky area and compress the image to 4096×1024 to save graphics memory.[11]

To understand the numbers in the example, consider that in a panorama image of 4096×2048 pixels, 1024 pixels represent 90°, 512px = 45°, 256px = 22.5°, 128px = 11.25°. To keep a top line of 11.25°, we keep an image height of $1024 + 128 = 1152$px, but the crop starts at pixel $Y = 1024 - 128 = 896$.

```
convert landscape.png -crop 4096x1152+0+896
        -resize 4096x1024! landscape_cropped.png
```

Note the exclamation mark in the -resize argument, which is required to stretch the image in a non-proportional way.

Alternatively, you can operate with **IM**'s "gravity", which indicates the corner or edge geometric offsets are referred to. Given that we want the lower part of the image to exist completely, you only need to compute the size of the cropped image:

```
convert landscape.png -gravity SouthWest -crop 4096x1152+0+0
        -resize 4096x1024! landscape_cropped.png
```

You still need the addition +0+0 in the -crop option, else the image will be cut into several pieces. In the file landscape.ini, you then have to set maptex_top=11.25.

---

[9]http://www.imagemagick.org/

[10]These may typically be .BAT files on Windows, or various shell scripts on Linux or Mac.

[11]Most modern graphics cards no longer require the "powers of two" image sizes, but we keep this practice to increase compatibility.

Command-line magick for old_style landscapes

Let us assume we want to create a high-resolution landscape from a pano image of width 16384 which we have carefully aligned and calibrated on an oversized grid template that also shows a measured horizon line (see 7.3.3). Usually it is not necessary to create the full-size image, but only the horizon range, in this high resolution. Assume this image has been aligned and justified on our grid image and is HEIGHT pixels high, the left border is at pixel X_LEFT, and top border (i.e., the point where relevant content like the highest tree is visible) is on pixel Y_TOP. Assume our graphics card is a bit oldish or you aim for maximum compatibility, so we can load only textures of at most 2048 pixels in size. Given that the horizon area usually only covers a few degrees, a vertical extent of 2048px seem a pretty good range for that most interesting zone. The ground can then be filled with some low-resolution image of grass, soil, or a properly oriented site map, or you can use **Hugin** to create a ground image (and using the maximum of 2048×2048 also here usually is far more than enough).

In **GIMP** (or **Photoshop**, ...), we must find the values for X_LEFT, Y_TOP and HEIGHT. HEIGHT is being resized to 2048, strictly, by the exclamation mark in the resize command. We can create our image tiles now with this singular beast of a command line (write all in 1 line!), which puts our files directly into STELLARIUM_LANDSCAPEPATH/LANDSCAPE_NAME:

```
convert PANO.png  -crop 16384xHEIGHT+X_LEFT+Y_TOP +repage
    -resize 16384x2048!
    -type TrueColorMatte -depth 8
    -crop 2048x2048 +repage
     png:STELLARIUM_LANDSCAPEPATH/LANDSCAPE_NAME/Horiz-%d.png
```

This creates 8 images. See section 7.1.4 for the landscape.ini where these images can be referenced. Don't forget to read off top and bottom lines (altitudes in degrees) from your grid, the vertical extent will form the decor_alt_angle, and the bottom line the decor_angle_shift entries in this file.

Creating a ground image for old_style landscapes

When you want a good ground image for an old_style landscape from your panorama and not just fill the groundtex with a monochrome texture or a map, you have to create a ground view in **Hugin**. But you may have already created a huge pano! This can also be used as source image, and a ground shot can be extracted with a reversed operation. In principle, all you need to know is the field of view around the nadir. Figure 7.5 shows a simple configuration file.

```
# hugin project file
#hugin_ptoversion 2
p f0 w2048 h2048 v92 E0 R0 n"TIFF_m⎵c:LZW⎵r:CROP"
m g1 i0 f0 m2 p0.00784314

# image lines
#-hugin  cropFactor=1
i w16384 h8192 f4 v360 Ra0 Rb0 Rc0 Rd0 Re0 Eev0 Er1 Eb1 r0
   p90 y0 TrX0 TrY0 TrZ0 Tpy0 Tpp0 j0 a0 b0 c0 d0 e0 g0 t0
   Va1 Vb0 Vc0 Vd0 Vx0 Vy0  Vm5 n"Eqirect_Pano360.png"
```

Figure 7.5: Project file ground.pto usable to create the ground image with **Hugin** or, on the command line, its **nona** stitcher. The last line, starting with i, has been wrapped, but must be 1 line.

Say, the side panels extend down to decor_angle_shift=-44 degrees, which means you must close the ground with a Nadir $FoV = 2 \times (90 - 44) = 92$. For maximum compatibility, we will again make an image of width and height both 2048px. These values can be found in the p line in Figure 7.5. The i line describes the input image, which is our full equirectangular pano of width w= 16384 and height h= 8192. The last argument of that line is the image file name.

For processing, we do not use the **Hugin** GUI, but simply the command line. The actual program to call is **nona**. If your stitched panorama is a 16-bit TIFF, **nona** will also make a 16-bit image, but our textures are limited to 8-bit PNGs. We apply our most useful tool, **convert** from the **ImageMagick** suite.

```
nona -v -m PNG  ground.pto -o ground.png
convert ground.png -depth 8 ground_8bit.png
```

The file ground_8bit.png is then used in the groundtex field on landscape.ini.

7.3.3   Final Calibration

The creation of a *calibrated panorama* (which can be regarded as dependable proxy for further measurements taken inside Stellarium) requires reference measurements to match the photos against. We must take azimuth/altitude measurements with a theodolite or total station, in the optimal case along the full horizon, and in addition I recommend to take azimuth and altitudes of some distinct features along the horizon which must also be visible in the photographs: mountain summits, electrical towers, church towers, ...

I recommend you create grid templates of the sizes you are going to create, e.g. 4096, 8192, 16386 and 20480 pixels wide with some diagram tool. On these, you can then also draw the measured horizon line.

Now, load a panorama on top of this in the **GIMP**, i.e., copy it into a separate layer over the grid image, and set it semi-transparent.

Try to align the center of the image (where the geometric anchor has been defined; remember: this should be the image pointing south!) with the measured horizon line or the distinct features.

The optimal solution consists of a photo panorama which aligns perfectly with the measured line and features. We now have to iteratively bring deviations to a minimum. The process depends on processor speed, image size, your training and – most of all – your requirements in accuracy!

In the **GIMP**, load your grid image with horizon line. Now select File ⟩ Open as Layers... , load your photo panorama, and then set layer transparency in the Layers dialog to about 50%.

Select the double-arrow tool to move the panorama via mouse drag and cursor keys over the grid, and align the outline of the photo horizon's southern point with the measured line. Now it's time to estimate the quality of the panorama.

In **Hugin**'s Photos tab, select the Positions view on the right side. Now you see "Yaw", "Pitch" and "Roll" values of camera-to-world orientation listed in the photos list. It should now be possible, by changing the values *only for the anchor image* and re-optimizing, to come to a panorama with only minimal error. In the process, start with Optimising Positions ⟩ incremental from anchor , then go for view and barrel optimisation, and so on. Always try to remove foreground match points which have large error and are irrelevant for the task to match the horizon. Those are especially cross-matches of horizon and subhorizon rows of images. Only vertically and horizontally adjacent images should be required to match. For handheld panoramas, also links between adjacent images in the non-horizontal rows are usually too erroneous to be useful, just remove these match points. Use the Layout tab in the Fast Panorama Preview to see the relations between images (Fig. 7.6): Red lines have big errors, green lines are good, thin gray lines indicate possible overlap without specified match points. After each optimisation step, export a new pano image, load as layer in **GIMP**, and check again.

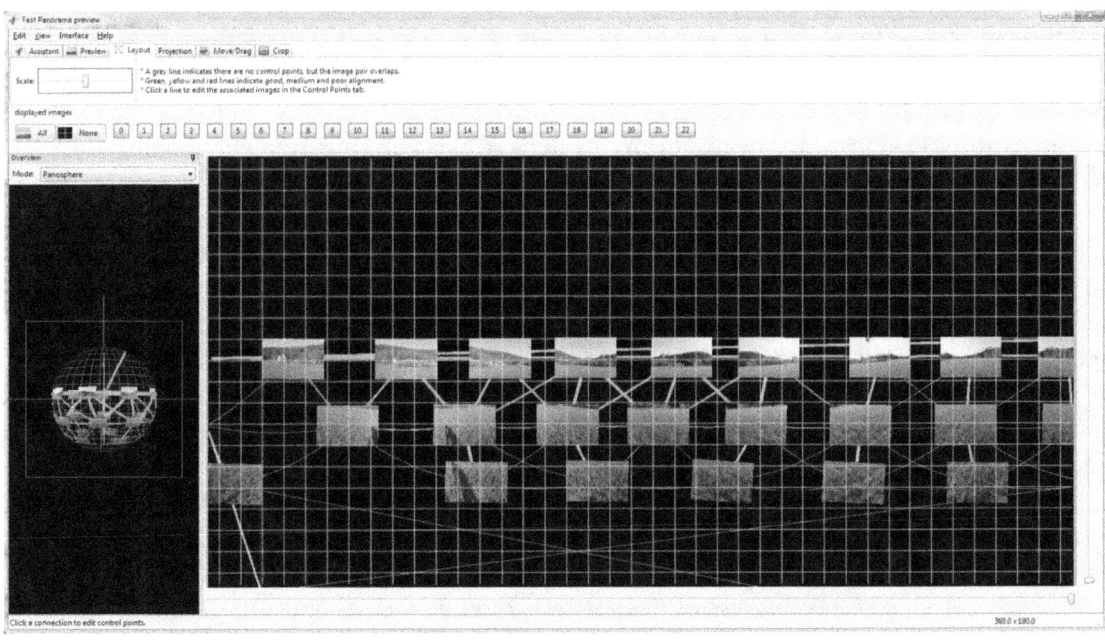

Figure 7.6: **Hugin**'s Fast Panorama Preview can be used to check which images are connected to its neighbours. Most important are good matches along the horizon, the images in the lower rows are clearly less important. If captured on a tripod, they should still match.

Basic rules to observe (use obvious inverses).
- If image aligns well in azimuth but overshoots the grid to the right: Increase yaw accordingly (0.022°/pixel if image is 16384 pixels wide).
- If the north end (left and right borders) is higher than the southern contact point: Increase pitch angle.
- If north and south points are OK, but the western (right) half is higher than the eastern (left) half: Increase Roll angle.

The corrections required for pitch and roll may be surprisingly small!

Within a few rounds of adjustments, panorama creation, adding as layer in the image editor, and comparing to the reference data, you should achieve a match to fit your needs.

In case you have taken photographs in several rings but without a panorama tripod, you may have to first align only the horizontal images (deselect the lower images to exclude from optimisation), and when the horizon ring is aligned perfectly, deactivate further optimisation in **Hugin** for those photos while "attaching" (optimising) the lower photos. In **Hugin**'s Photos tab, select Optimise ⟩ Geometric ⟩ Custom Parameters. This opens an extra tab Optimiser, where you can fine-tune your needs: Switch off all variables for the photos in the horizon ring, and make sure the lower photos fit in the preview after optimisation.

It may even help to define that the lower rows have been taken with a different Lens, so the field of view and distortion settings of the horizon row will be used as it had been found during the horizon-only match.

By now you should have enough experience what level of error may be acceptable for you.

7.3.4 Artificial Panoramas

I have created a website[12] where you can enter geographical coordinates and download a file pano.kml which helps with image creation from **Google Earth** imagery. Store this file for a site, let us call it *MYPLACE*, into a new directory GE_MYPLACE inside your landscapes directory.

Store all scenes visible from the respective viewpoint *MYPLACE* as picture into one common folder in your landscapes/GE_MYPLACE under the viewpoint name, e.g., 75-30.jpg, which means 75 degrees from Nadir, azimuth 30 degrees. Also, double-click the pano entry or the marker in **Google Earth** to open a window with the basic content of your landscape.ini. Copy and paste from there into a new file landscape.ini and adjust the obvious entries. Complete as required with the entries described in section 7.1.3.

On loading of the images, Hugin will not be able to detect any EXIF lens data and ask you for the horizontal field of view. Enter 60 degrees, which is the standard value for **Google Earth** screenshots[13].

The viewpoint names translate almost directly to the yaw and pitch angles which you can enter in the image list in **Hugin**'s Photos tab. For example, switch to the Positions display on the right window edge in the Photo tab, mark all images that start with 25- and assign a pitch angle of $-90 + 25 = -65$. The second part of the names is directly the azimuth. In this case, don't run the optimizer, but you can immediately set an output resolution and stitch (see 7.2.3). To get rid of the image decorations (compass etc), apply masks[14]. Postprocessing steps are the same as for photo-panoramas: make sky invisible, crop, etc.

It is also interesting to switch on the 3D buildings layer before creating the images. If temples or other buildings are accurate, this will give an even closer approximation to what would be visible on-site. Note however that not every building will be modelled in usable quality, and that usually vegetation is not included in the 3D buildings layer. Also, if you are too close to buildings, they may be cut away by the *near clipping plane* of the rendering.

These images, based on **Google Earth** imagery and the SRTM topographic model, seem usable as *first rough approximation* to a photo-based or surveyed panorama. Note that it is definitely not accurate enough for representing nearby horizon features or critically important mountain peaks, and please note that Google has image copyright which at least requires you to acknowledge when displaying these pictures.

7.3.5 Nightscape Layer

Since version 0.13, Stellarium can simulate artificial illumination, like streetlamps, bright windows, or the skyglow over cities[76]. One way to create this layer is to make 2 panorama series during the day and night and process these in the same **Hugin** project to align those photos, and then stitch two separate images by selecting either the daylight or the nighttime shots. The night panorama has to be processed to remove stars, airplanes, etc.

The other way is a simple layer overpainted in the image processing program. As rough recommendation, use several layers to prepare this feature:

- Put a semitransparent black layer over your daylight image, this helps you to place your painted pixels.
- Paint windows, street lamps, signs, ... You may apply a layer style to produce some glow.
- To draw an impression of more light in the atmosphere (city skyglow), use a gradient with some brownish color. Generally the color depends on the appropriate mix of city lights (sodium, mercury vapour, etc.). Note that on the city outskirts a simple vertical gradient will

---

[12]http://homepage.univie.ac.at/Georg.Zotti/php/panoCam.php
[13]Note that if you work with **Google Earth Pro**, you can create different FoV!
[14]There is a wide overlap in the images to allow generous trimming.

not work, towards the city the horizon is much brighter. Use a huge but weak brush to make a more spotty sky.

- Use the existing landscape as template for the layer mask for this gradient sky layer. (You want to hide skyglow by leaves in the foreground!)
- If you want to add only a few lights to an old_style landscape, you need to provide only the panels showing those lights. Just load a side panel for reference, place a new layer on top, and paint the lights on windows, lamps etc. There is no light option for the ground texture. This makes old_style landscapes best suited for localized light pollution, not city skyglow.

The resulting image is then declared in the maptex_illum line of landscape.ini. Try also to balance the global strength of light pollution with the light_pollution key, and a probable minimal brightness with the minimal_brightness key.

Try to match the visual appearance, not necessarily what photographs may have recorded. E.g., the *Grossmugl* sky shows horizon glow mostly towards the city of Vienna, where long-time exposures may already be saturated.

The possibilities seem limited only by your time and skills!

7.4 Other recommended software

Here is a short collection of other useful programs for (panorama) image manipulation and other tasks on Windows.

7.4.1 IrfanView

IrfanView is a free image viewer for Windows with many options. It can show almost any image format, including several camera RAW formats, in windowed and full-screen mode. It is definitely preferrable over any image viewer built into Windows. Unfortunately however, it has no panorama viewer function!

7.4.2 FSPViewer

FSPViewer[15] by Fulvio Senore is an excellent panorama viewer for equirectanglar images. Images centered along the horizon can be viewed directly, while settings for images with different minimum and maximum angles, as well as "hotspots" (similar to hyperlinks) which move to neighboring panoramas, can be configured in an .FSV text file like figure 7.7.

```
ImageName=Horizon_Rosenburg.jpg
WindowTitle=Horizon_Rosenburg
hFov=70
#Formula:  HP=100*(h/2-upper)/(lower-upper) in Hugin crop, or
#          HP=100*zeroRow/imgHeight
HorizonPosition=33.8
```

Figure 7.7: FSP configuration file (example)

7.4.3 Clink

Clink[16] is a command line enhancement for Windows developed by Martin Ridgers. If you have ever worked under a Linux **bash**-like command line, you will easily feel that Windows' **cmd.exe**

---

[15]Further details are available on its home page http://www.fsoft.it/FSPViewer/.

[16]http://mridgers.github.io/clink/

is extremely limited. **Clink** provides several useful features, most notably a really usable command-line completion. It is not essential for our tasks, but a general improvement of usability of the Windows command line which else has not caused me any trouble.

7.4.4 Cygwin

Compared to Linux, the command line of Windows can be a humbling experience. None of the wonderful helpers taken for granted on Linux are available. **Cygwin**[17] provides a command line console with **bash** shell and all the niceties like **make**, **awk**, **sed**, etc. which seem essential for routine work. If you are used to Linux tools, use inline scripts in your Makefiles and need more than **Clink** can offer, you should install **Cygwin**.

7.4.5 GNUWin32

Alternative to **Cygwin**, several of those nice tools (**sed**, **awk** etc.) have also been made available as standalone commands for Windows. If you don't need the inline scripting capabilities in Makefiles which you would get from **Cygwin** but just want to call **awk** or **sed** inside your **.BAT** scripts, maybe this is enough.

---

[17]https://cygwin.com/index.html

8. Deep-Sky Objects

Since version 0.10.0 Stellarium uses the "json" cataloguing system of configuring textures. At the same time the Simbad online catalogue was added to the search feature, making the catalog somewhat redundant and used now only as a first search point or if there is no internet connection.

If the object has a name (not just a catalogue number), you should add one or more records to the .../nebulae/default/names.dat file (where ... is either the installation directory or the user directory). See section 8.1.2 Modifying names.dat for details of the file format.

If you wish to associate a texture (image) with the object, you must add a record to the .../nebulae/default/textures.json file. See section 8.1.3 for details of the file format.

8.1 Stellarium DSO Catalog

Stellarium's DSO Catalog contains over 14000 objects and is available for end users as collection of files:

catalog.txt	Stellarium DSO Catalog in ASCII format for editing data
catalog.dat	Stellarium DSO Catalog in binary format for usage within Stellarium
names.dat	List of proper names of the objects from file catalog.dat

ASCII file can be converted into binary format through enabling an option in the file config.ini (See 5.3):

```
[devel]
convert_dso_catalog = true
```

The file catalog.txt should be put into the directory .../nebulae/default/.

Stellarium DSO Catalog contains data and supports the designations for follow catalogues:

NGC New General Catalogue

IC Index Catalogue

M Messier Catalog

C Caldwell Catalogue
B Barnard Catalogue [5]
Sh2 Sharpless Catalogue [56]
VdB Van den Bergh Catalogue of reflection nebulae [65]
RCW A catalogue of $H\alpha$-emission regions in the southern Milky Way [51]
LDN Lynds' Catalogue of Dark Nebulae [29]
LBN Lynds' Catalogue of Bright Nebulae [30]
Cr Collinder Catalogue [16]
Mel Melotte Catalogue of Deep Sky Objects [40]
PGC HYPERLEDA. I. Catalog of galaxies[1]
UGC The Uppsala General Catalogue of Galaxies
Ced Cederblad Catalog of bright diffuse Galactic nebulae [13]

Cross-index data for Stellarium DSO Catalog is partially obtained from "Merged catalogue of reflection nebulae" [31] and astronomical database SIMBAD [68].

8.1.1 Modifying catalog.dat

This section describes the inner structure of the files `catalog.dat` (binary format) and `catalog.txt` (ASCII format). Stellarium can convert ASCII file into the binary format file for faster usage within the program.

Each line contains one record, each record consisting of the following fields with *tab* char as delimiter:

Column	Type	Description
1	integer	Deep-Sky Object Identificator
2	float	RA (decimal degrees)
3	float	Dec (decimal degrees)
4	float	B magnitude
5	float	V magnitude
6	string	Object type (See section 8.1.1 for details).
7	string	Morphological type of object
8	float	Major axis size or radius (arcmin)
9	float	Minor axis size (arcmin)
10	integer	Orientation angle (degrees)
11	float	Redshift
12	float	Error of redshift
13	float	Parallax (mas)
14	float	Error of parallax (mas)
15	float	Non-redshift distance (Mpc for galaxies, kpc for other objects)
16	float	Error of non-redsift distance (Mpc for galaxies, kpc for other objects)
17	integer	NGC number (New General Catalogue)
18	integer	IC number (Index Catalogue)
19	integer	M number (Messier Catalog)
20	integer	C number (Caldwell Catalogue)
21	integer	B number (Barnard Catalogue)
22	integer	Sh2 number (Sharpless Catalogue)
23	integer	VdB number (van den Bergh Catalogue of reflection nebulae)
24	integer	RCW number (A catalogue of $H\alpha$-emission regions in the southern Milky Way)

[1]The PGC and UGC catalogues have a partial support

25	integer	LDN number (Lynds' Catalogue of Dark Nebulae)
26	integer	LBN number (Lynds' Catalogue of Bright Nebulae)
27	integer	Cr number (Collinder Catalogue)
28	integer	Mel number (Melotte Catalogue of Deep Sky Objects)
29	integer	PGC number (HYPERLEDA. I. Catalog of galaxies); partial
30	integer	UGC number (The Uppsala General Catalogue of Galaxies); partial
31	string	Ced number (Cederblad Catalog of bright diffuse Galactic nebulae)

Types of Objects

Possible values for type of objects in the file `catalog.dat`.

Type	Description
G	Galaxy
GX	Galaxy
AGX	Active Galaxy
RG	Radio Galaxy
IG	Interacting Galaxy
GC	Globular Cluster
OC	Open Cluster
NB	Nebula
PN	Planetary Nebula
DN	Dark Nebula
RN	Reflection Nebula
C+N	Cluster associated with nebulosity
HII	HII Region
SNR	Supernova Remnant
BN	Bipolar Nebula
EN	Emission Nebula
SA	Stellar Association
SC	Star Cloud
CL	Cluster
IR	Infra-Red Object
QSO	Quasar
Q?	Possible Quasar
ISM	Interstellar Matter
EMO	Emission Object
LIN	LINEAR-type Active Galaxies
BLL	BL Lac Object
BLA	Blazar
MOC	Molecular Cloud
YSO	Young Stellar Object
PN?	Possible Planetary Nebula
PPN	Protoplanetary Nebula
*	Star
**	Double Star
MUL	Multiple Star
empty	Unknown type, catalog errors, *Unidentified Southern Objects* etc.

8.1.2 Modifying names.dat

Each line in the file names.dat contains one record. A record relates an extended object catalogue number (from catalog.dat) with a name. A single catalogue number may have more than one record in this file.

The record structure is as follows:

Offset	Length	Type	Description
0	5	%5s	Designator for catalogue (prefix)
5	15	%d	Identificator for object in the catalog
20	60	%s	Proper name of the object (translatable)

If an object has more than one record in the file names.dat, the last record in the file will be used for the nebula label.

8.1.3 Modifying textures.json

This file is used to describe each nebula image. The file structure follows the JSON format, a detailed description of which may be found at www.json.org. The textures.json file which ships with Stellarium has the following structure:

serverCredits (optional) a structure containing the following key/value pairs:

> **short** a short identifier of a server where the json file is found, e.g. "ESO"
>
> **full** a longer description of a server, e.g. "ESO Online Digitised Sky Survey Server"
>
> **infoURL** a URL pointing at a page with information about the server

imageCredits a structure containing the same parts as a serverCredits structure but referring to the image data itself

shortName an identifier for the set of images, to be used inside Stellarium

minResolution minimum resolution, applies to all images in the set, unless otherwise specified at the image level

maxBrightness the maximum brightness of an image, applies to all images in the set, unless otherwise specified at the image level

subTiles a list of structures describing indiviual image tiles, or referring to another json file. Each subTile may contain:

> **minResolution**
>
> **maxBrightness**
>
> **worldCoords**
>
> **subTiles**
>
> **imageCredits**
>
> **imageUrl**
>
> **textureCoords**

shortName (name for the whole set of images, e.g. "Nebulae")

miniResolution (applies to all images in set)

alphaBlend (applies to all images in set)

subTiles list of images. Each image record has the following properties:

> **imageCredits** (itself a list of key/pairs)
>
> **imageUrl** (e.g. file name)
>
> **worldCoords** (a list of four pairs of coordinates representing the corners of the image)
>
> **textureCoords** (a list of four pairs of corner descriptions. i.e. which is top left of image etc)
>
> **minResolution** (over-rides file-level setting)

maxBrightness

Items enclosed in Quotation marks are strings for use in the program. Syntax is extremely important. Look at the file with a text editor to see the format. Items in <> are user provided strings and values to suit the texture and source.

```
{
   "imageCredits"   : { "short" : "<author␣name>" ,
                        "infoUrl" : "http://<mysite.org>"
                     },
   "imageUrl"       : "<myPhoto.png>",
   "worldCoords"    : [[[ X0, Y0], [ X1, Y1], [ X2, Y2], [ X3, Y3] ]],
   "textureCoords"  : [[[ 0,0],[1,0],[1,1],[0,1]]],
   "MinResolution"  : 0.2148810463,
   "maxBrightness"  : <mag>
},
```

where

worldCoords Decimal numerical values of the J2000 coordinates (RA and dec both in degrees) of the corners of the texture. These values are usually given to 4 decimal places.

textureCoords Where 0,0 is South Left, 1,0 the South Right, 1,1 North Right, 0,1 North Left corners of the texture.

MinResolution UNDOCUMENTED VALUE! Sorry!

maxBrightness total object brightness, magnitude

Calculating of the coords of the corners of the images (plate solving) is a time consuming project and needs to be fine tuned from the screen display. As most images will be two dimensional, display on a spherical display will limit the size to about 1 degree before distortion becomes evident. Larger images should be sectioned into a mosaic of smaller textures for a more accurate display.

8.2 Adding Extra Nebulae Images

BARRY GERDES

8.2.1 Preparing a photo for inclusion to the `textures.json` file

Figure 8.1: Screen shot of nebula images displayed in Stellarium

The first step is to take a photo of the object you wish to display in Stellarium. When you have the picture you will need to align it with the equatorial coordinate system so that north is directly up and not inverted side to side or up and down as can happen with photos taken with a diagonal mirror in the path. Next you will need to crop the picture, setting the main feature at the centre and making the cropped size a factor of 2^n eg. 64, 128, 256, 512, 1024 or 2048 pixels square (or elongated like 512x1024). If this requirement is not met, your textures may not be visible, or graphics performance may be seriously impacted. Textures larger than 2048 may only be supported on high-end hardware. Images must be in PNG format. When cropping, make sure you leave at least six prominent background stars.

The next step is to process your photo to make the background black, really black. This will ensure that your background will meld with the Stellarium background and not be noticed as gray square. Suitable programs to do all this are **The GIMP**[2] or **Photoshop** if you can afford it.

When you have your image prepared you will need to plate solve it using at least 6 known GSC stars that can be identified. That is why the cropping with plenty of stars was necessary. When the plate is solved you will need to find the J2000 coordinates of the corners and convert them to decimal values to form the world coordinates in the `textures.json` file.

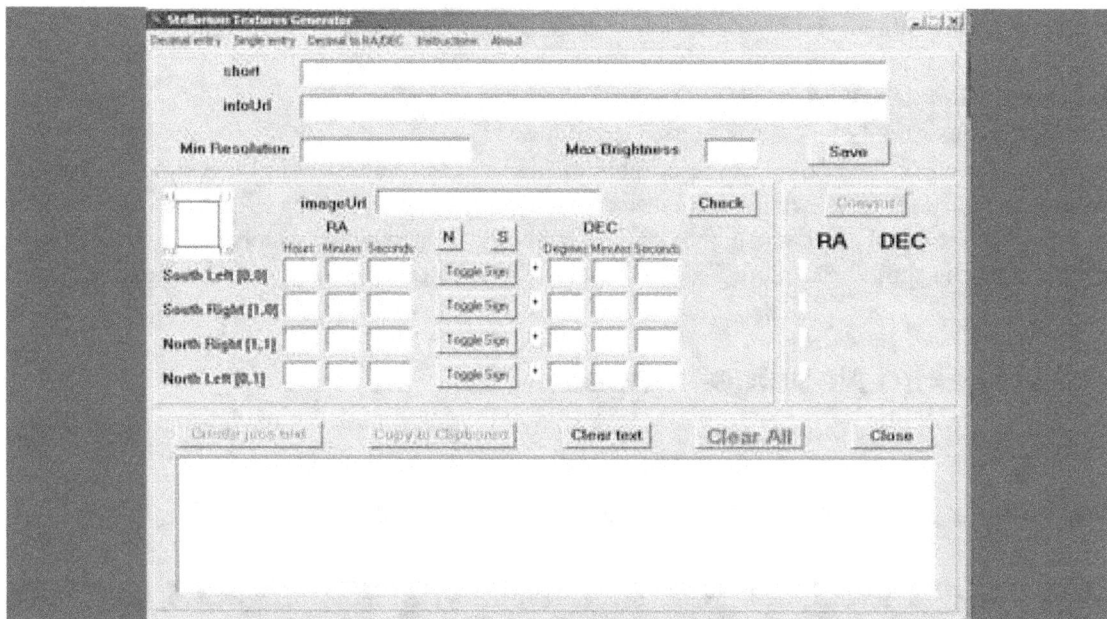

Figure 8.2: **Stellarium Textures Generator**: A program to convert Equatorial coordinates into decimal form and write a `textures.json` insert

The program **Stellarium Textures Generator** by Peter Vasey (Fig. 8.2) can convert the corner coordinates of a texture found in your plate solving program into decimal values and write an insert for the `textures.json` file.[3]

There is another program, **ReadDSS** (Fig. 8.3), written by Barry Gerdes in Qb64(gl), that will perform the same task but allows manipulation of the epochs.[4]

---

[2]free in keeping with the Stellarium spirit; available from `http://www.gimp.org`

[3]It is available as a freebee from `http://www.madpc.co.uk/~peterv/astroplover/equipnbits/Stellariumtextures.zip`.

[4] `http://barry.sarcasmogerdes.com/stellarium/uploads/writejsoninsert.zip`

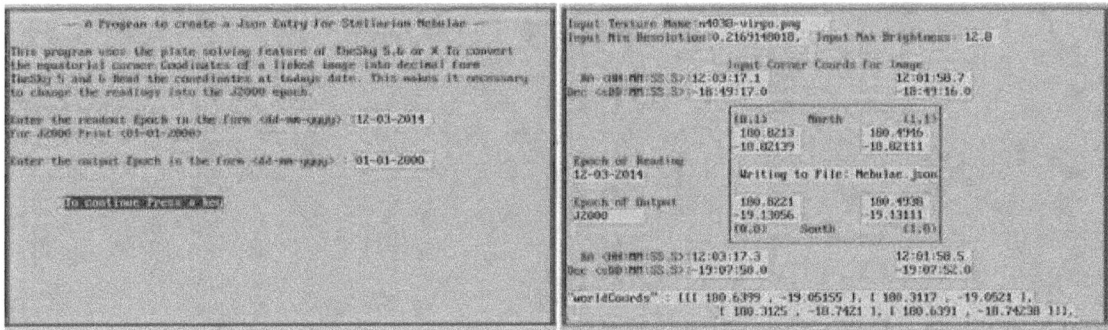

Figure 8.3: **ReadDSS**: A program to write a `textures.json` insert with epoch manipulation.

8.2.2 Plate Solving

Suitable programs that can accept your picture and calculate its corner coordinates are hard to find. I have only found one that suits our purpose and it is another expensive planetarium program, **TheSky X Pro**. However the older versions **TheSky 5** and **6 Pro** will also do the job if suitably configured, although I could not solve the test program with **TheSky 6** that uses the same procedure as **TheSky 5**.

These programs have a link feature that can match your photo to the selected area of the screen and superimpose it on the display with a box around your photo provided it can match at least 6 stars from the GSC that is included with the program. When this is fitted you can read the corner coordinates of your texture in the Status bar by selecting them with a mouse. **TheSky X** can read these coordinates in J2000 values and uses textures in the FITS format, but the earlier programs only read the coordinates of the current program date. To read the J2000 coordinates it is necessary to re-start the program with the date set to 1-1-2000.

To add the picture to **TheSky 5** you need first make a mono 8 bit version of the photo and place it on the clipboard. Run **TheSky** and centre on the object centre. Look in the Tools menu for the image link and select setup. Tick show image frame to put a frame around the image.

Paste the clipboard image on the display and use the zoom and position controls to get it as close to the size and position as possible by visually matching stars. Go to the menu again and click on link wizard. If you have been successful the window will show the number of stars matched and the option to accept or continue. Accept and you will now see all the matched stars have overlaid the picture. You can now read off the corner coordinates from the status bar starting at the bottom (south) left and continuing counterclockwise to the top (north) left.

8.2.3 Processing into a `textures.json` insert

Place your image in the `*.png` format in the `.../nebulae/default/` folder. Ensure that the name matches the `textures.json` entry.

Once you have the corner coordinates of your photo you can add them to the decimal converter program and it will write an insert `nebula.json` as a text file that you can paste directly into the `textures.json` file that is in the `.../nebulae/default/` folder.

Save the `textures.json` file with the new insert and run Stellarium. Find the object in the F3 Object selection window and slew to it. Your image should be there and with a bit of luck it will nicely overlay the stars in Stellarium. However this only rarely happens, so a little bit of tweaking of the JSON `worldcoords` will be needed to get a perfect match. Select equatorial mode (✳ or Ctrl + M). This will show the area with north up. Select each corner in sequence and make small changes to the coordinates. Restart Stellarium each time and check if you have moved into the right direction. Continue with each corner until all the stars match. With a little bit of practice this will be done in about 10 minutes.

9. Adding Sky Cultures

GEORG ZOTTI

Stellarium comes with a nice set of skycultures. For ethnographers or historians of science it may be a worthwile consideration to illustrate the sky culture of the people they are studying. It is not very hard to do so, but depending on your data, may require some skills in image processing.

Some features regarding translation and multilinguality have evolved over the years, and not all skycultures currently included in Stellarium adhere to the standards described in the following sections. If you add a new skyculture, please do so for an optimal result!

In the Stellarium program folder you can see a folder `skycultures`. Let us assume you work on Windows and want to create a new Skyculture, say, *myCulture*.

You can take the `inuit` directory as template to start with. Just copy the folder `C:\Program Files\Stellarium\skycultures\inuit` to `C:\Users\[YOU]\AppData\Roaming\Stellarium\skycultures\myculture`

In the folder you see image files for the constellation artwork, and all other files with various extensions are text files.

9.1 Basic Information

In `myculture\info.ini`, change the entries to

```
[info]
name=myCulture
author=me
```

(or what seems best for you). The name is used for the list entry in the Starlore tab in the View dialog (see 4.4.5).

9.2 Skyculture Description Files

In order to have translated texts we have files description.<LANG>.utf8, where <LANG> is the two-letter ISO 639-1 language code, or its extension which contains language and country code, like pt_BR for Brazilian Portuguese. A minimum skyculture must contain the file description.en.utf8, this is en=english text with optional HTML tags for sections, tables, etc. You can also have embedded images in the HTML (your book cover? Views of sacred landscapes/buildings/artwork/. . . ?), just make them PNG format please. The length of the description texts is not limited, you have room for a good description, tables of names/translations, links to external resources, whatever seems suitable. When you started from a copied skyculture, delete the other description.*.utf8 files.

If you can provide other languages supported by Stellarium, you can provide translations yourself, else Stellarium translators *may* translate the English version for you. (It may take years though.) The file ending .utf8 indicates that for special characters like ÄÖÜßáé you should use UTF8 encoding. If you write only English/ASCII, this may not be relevant.

9.3 Constellation Names

The native constellations are listed in constellation_names.eng.fab. It consists of 3 simple columns: Abbreviation(or just a serial number), native name, and english translation. The writing _("name") allows automatic translation of the English strings to other languages. These strings will be used as constellation labels.

The first column (abbreviation) in the Western sky culture provides the canonical 3-letter abbreviation for constellations as used by the International Astronomical Union. Such abbreviations may not be available for the skyculture you are working with, so you must invent your own. These abbreviations are used as keys in the other files, so they must be unique within your skyculture. It is not necessary to have 3-letter keys.

The keys can be displayed on screen when labels are requested in the Starlore GUI (section 4.4.5). If you want to prevent certain abbreviations from being displayed, let them start with a dot. See the effect in the Western (H.A.Rey) skyculture: In Abbreviated mode, only the official abbreviations are displayed. In Native mode, the second column of constellation_names.eng.fab is shown. Only with setting Translated, the text translated from the text shown in the third column is shown. If your skyculture is a variant of the Western skyculture, please use the canonical Latin names, they have all been translated already.

If your skyculture is not a variant of the generally known Western skyculture, please include an English translation to the name given in the native language. Else translators will not be able to translate the name. See a good example in the Mongolian skyculture.

9.4 Star Names

The file star_names.fab contains a list of HIP catalogue numbers and common names for those stars. Each line of the file contains one record of two fields, separated by the pipe character (|). The first field is the Hipparcos catalogue number of the star, the second is the common name of the star in a format that enables translation support, e.g:

```
113368|_("Fomalhaut")
```

9.5 Planet Names

The file `planet_names.fab` contains a list of native names of planets. Each line of the file contains one record of 3 fields, separated by the white space or tab character. The first field is the English name of the planet, the second is the native name of the planet (can be in the native language, but please for maximum utability use an english transliteration) and the third is the translatable version of the native name of the planet (translated into English). Here is an example from the Egyptian skyculture:

```
Mars      "Horus-Desher"   _("Red␣Horus␣(Mars)")
```

9.6 Stick Figures

The modern-style stick figures are coded in `constellationship.fab`. Lines look like:

```
Abbr pairs pair1_star1 pair1_star2 pair2_star1 pair2_star2 ...
```

In this file,

Abbr is the abbreviation defined in `constellation_names.eng.fab`

pairs is the number of line pairs which follow.

pairN_starA Hipparcos numbers for the stars which form the constellation stick figure. We need two entries per line, longer line segments are not supported. To find the HIP number, just have Stellarium open and click on the star in Stellarium while editing this file.

9.7 Constellation Borders

The optional file `constellations_boundaries.dat` includes data for the border lines drawn between constellations. The western constellations have been given borders based on B1875.0 coordinates, and all skycultures with names starting in `Western_` use these borders automatically.

The format for this file is a bit more dificult than the other files. It contains sections which may consist of multiple lines, of the format:

```
N RA_1 DE_1 RA_2 DE_2 ... RA_N DE_N 2 CON1 CON2
```

where

N number of corners

RA_n, DE_n right ascension and declination (degrees) of the corners in J2000 coordinates.

2 CON1 CON2 legacy data. They indicated "border between 2 constellations, CON1, CON1" but are now only required to keep the format.

9.8 Constellation Artwork

Constellation artwork is optional, but may give your skyculture the final touch, if it requires artwork at all. E.g., H. A. Rey's variant of the Western skyculture deliberately does not contain artwork.

Each constellation artwork is linked to 3 stars in its constellation. This is programmed in the file `constellationsart.fab`. You have to write lines

```
Abbr image_name.png x1 y1 HIP1 x2 y2 HIP2 x3 y3 HIP3
```

where

Abbr is the abbreviation defined in `constellation_names.eng.fab`

image_name.png is the file name of your texture. It should be sized in a power of two, like 512×512, 1024×2048 etc. Avoid dimensions larger than 2048, they are not supported on all systems. You can distort images to better exploit the pixels, the texture will be stretched back. The background of the artwork image must be absolutely black.

xn, yn, HIPn pixel locations of the star in the constellation drawing (find those in any image editor) and HIPn is the star number in the Hipparcos catalog, which you find when you click on the star in Stellarium.

In case the artwork is only available in a certain projection (e.g., an all-sky map), or is otherwise heavily distorted so that the match is not satisfactory, you may have to reproject the image somehow. For aligning, you should switch Stellarium to Stereographic projection for optimal results.

You don't have to shutdown and restart Stellarium during creation/matching, just switch skyculture to something else and back to the new one to reload.

9.9 Seasonal Rules

File `seasonal_rules.fab` (optional) contains possible seasonal rules for the visibility of constellations. There is one rule per line. Each rule contains three elements separated with white space (or tab character): constellation ID, start of visibility (month) and end of visibility (month), e.g:

```
Emu 6 3
```

This specifies that constellation Emu (abbreviated also as "Emu") is visible only from June to March.

9.10 Publish Your Work

If you are willing to let other users enjoy the result of your hard work (and we certainly hope you do!), when you are done, please write a note in the Forum or at Launchpad. Please be prepared to put the imagery and text under some compatible open-source license (Creative Commons). Else the skyculture cannot be hosted by us.

10   Plugins 95
10.1   Enabling plugins
10.2   Data for plugins

11   Interface Extensions 97
11.1   Angle Measure Plugin
11.2   Compass Marks Plugin
11.3   Equation of Time Plugin
11.4   Field of View Plugin
11.5   Pointer Coordinates Plugin
11.6   Text User Interface
11.7   Remote Control
11.8   Solar System Editor Plugin
11.9   Timezone Configuration Plugin

12   Object Catalog Plugins 111
12.1   Bright Novae Plugin
12.2   Historical Supernovae Plugin
12.3   Exoplanets Plugin
12.4   Pulsars Plugin
12.5   Quasars Plugin
12.6   Meteor Showers Plugin
12.7   Navigational Stars Plugin
12.8   Satellites Plugin
12.9   ArchaeoLines Plugin

13   Scenery3d – 3D Landscapes 139
13.1   Introduction
13.2   Usage
13.3   Hardware Requirements & Performance
13.4   Model Configuration
13.5   Predefined views

14   Stellarium at the Telescope 151
14.1   Oculars Plugin
14.2   TelescopeControl Plugin
14.3   StellariumScope plugin
14.4   Other telescope servers and Stellarium
14.5   Observability Plugin

15   Scripting 161
15.1   Introduction
15.2   Script Console
15.3   Includes
15.4   Minimal Scripts
15.5   Example: Retrograde motion of Mars
15.6   More Examples

10. Plugins

Starting with version 0.10.3, Stellarium's packages have included a steadily growing number of plug-ins: Angle Measure, Compass Marks, Oculars, Telescope Control, Text User Interface, Satellites, Solar System Editor, Time Zone, Historical Supernovae, Quasars, Pulsars, Exoplanets, Observability analysis, ArchaeoLines, Scenery3D. All these plug-ins are "built-in" in the standard Stellarium distribution and DON'T need to be downloaded separately.

10.1 Enabling plugins

To enable a plugin:

1. Open the **Configuration dialog** (press F2 or use the left tool bar button)
2. Select the **Plugins** tab
3. Select the plugin you want to enable from the list
4. Check the **Load at startup** option
5. Restart Stellarium

If the plugin has configuration options, the **configuration** button will be enabled when the plugin has been loaded and clicking it will open the plugin's configuration dialog. When you only just activated loading of a plugin, you must restart Stellarium to access the plugin's configuration dialog.

10.2 Data for plugins

Some plugins contain files with different data, e.g., catalogs. JSON is a typical format for those files and you can edit its content manually. Of course, each plugin has a specific format of data for the own catalogs, and you should read documentation for the plugin before editing of its catalog.

You can read some common instructions for editing catalogs of plugins below. In this example we use file name `catalog.json` for identification of catalog for a typical plugin.

You can modify the `catalog.json` files manually using a text editor. **If you are using**

Windows, it is strongly recommended to use an advanced text editor such as Notepad++[1] to avoid problems with end-of-line characters. (It will also color the JSON code and make it easier to read.)

Warning: Before editing your `catalog.json` file, make a backup copy. Leaving out the smallest detail (such as a comma or forgetting to close a curly bracket) will prevent Stellarium from starting.

As stated in section 5, the path to the directory[2] which contains `catalog.json` file is something like:

Windows C:\Users**UserName**\AppData\Roaming\Stellarium\modules*PluginName*

Mac OS X **HomeDirectory**/Library/Preferences/Stellarium/modules/*PluginName*

Linux and UNIX-like OS ~/.stellarium/modules/*PluginName*

---

[1]`http://notepad-plus-plus.org/`

[2]This is a hidden folder, so in order to find it you may need to change your computer's settings to display hidden files and folders.

11. Interface Extensions

Most users will soon be familiar with the usual user interface. A few plugins are available which extend the regular user interface with a few small additions which are presented first. However, some applications and installations of Stellarium require completely different user interfaces. Mostly, these serve to avoid showing the user interface panels to an audience, be that in your astronomy club presentations, a domed planetarium or in a museum installation.

11.1 Angle Measure Plugin

goes misty eyed
I recall measuring the size of the Cassini Division when I was a student. It was not the high academic glamor one might expect...It was cloudy...It was rainy...The observatory lab had some old scopes set up at one end, pointing at a *photograph* of Saturn at the other end of the lab. We measured. We calculated. We wished we were in Hawaii. A picture is worth a thousand words.

The Angle Measure plugin is a small tool which is used to measure the angular distance between two points on the sky.

1. Enable the tool by clicking the tool-bar button, or by pressing Ctrl + A . A message will appear at the bottom of the screen to tell you that the tool is active.
2. Drag a line from the first point to the second point using the left mouse button
3. To clear the measurement, click the right mouse button
4. To deactivate the angle measure tool, press the tool-bar button again, or press Ctrl + A on the keyboard.

In the configuration dialog, you can configure if you want to have distances given on the rotating sphere, or in horizontal (alt-azimuthal) coordinates. You can also link one point to the resting horizon, the other to the sky and observe how angles change.

11.2 Compass Marks Plugin

Stellarium helps the user get their bearings using the cardinal point feature – the North, South, East and West markers on the horizon. Compass Marks takes this idea and extends it to add markings every few degrees along the horizon, and includes compass bearing values in degrees.

When activated (see section 10.1), there is a tool bar button ⬚ for toggling the compass markings. Note that when you enable compass marks, the cardinal points will be turned off.

11.3 Equation of Time Plugin

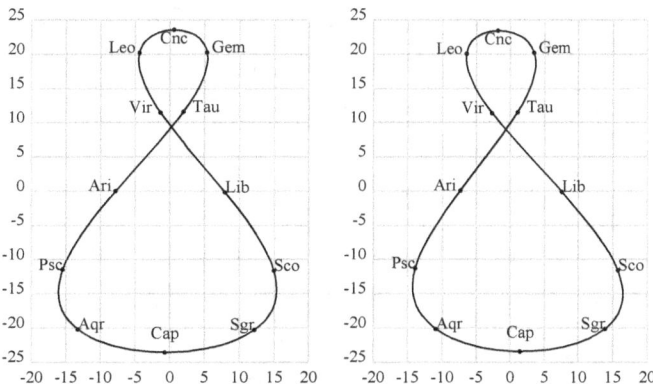

Figure 11.1: Figure-8 plots for Equation of Time, for years 1000 (left) and 2000 (right). These plots, often found on sundials, link solar declination (vertical axis) and its deviation at mean noon from the meridian, in minutes. Labeled dots indicate when the sun entered the respective Zodiacal sign (30°section of the ecliptic). Figures by Georg Zotti.

The Equation of Time plugin shows the solution of the equation of time. This describes the discrepancy between two kinds of solar time:

Apparent solar time directly tracks the motion of the sun. Most sundials show this time.

Mean solar time tracks a fictitious "mean" sun with noons 24 hours apart.

There is no universally accepted definition of the sign of the equation of time. Some publications show it as positive when a sundial is ahead of a clock; others when the clock is ahead of the sundial. In the English-speaking world, the former usage is the more common, but is not always followed. Anyone who makes use of a published table or graph should first check its sign usage.

If enabled (see section 10.1), click on the Equation of Time button on the bottom toolbar to display the value for the equation of time on top of the screen.

11.3.1 Section [EquationOfTime] in config.ini file

You can edit config.ini file by yourself for changes of the settings for the Equation of Time plugin – just make it carefully!

ID	Type	Description
enable_at_startup	bool	Display solution of the equation of time at startup of the planetarium
flag_use_ms_format	bool	Set format for the displayed solution - minutes and seconds and decimal minutes
flag_use_inverted_value	bool	Change sign of the equation of time
flag_show_button	bool	Enable displaying plugin button on the bottom toolbar
text_color	R,G,B	Color of font for the displayed solution of the equation of time
font_size	int	Font size for the displayed solution of the equation of time

11.4   Field of View Plugin

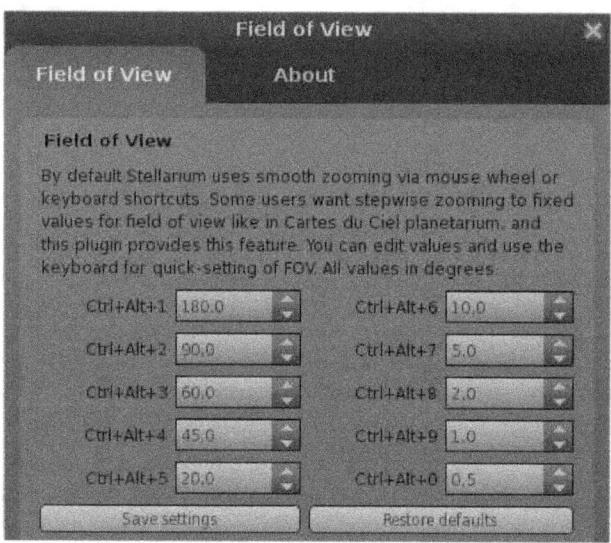

Figure 11.2: Configuration dialog of Field of View plugin

By default Stellarium uses smooth zooming via mouse wheel or keyboard shortcuts. Some users may want stepwise zooming to fixed values for field of view like in the **Cartes du Ciel**[1] planetarium program, and this plugin provides this feature. You can edit values and use the keyboard for quick-setting of FOV. All values in degrees.

1. Enable the tool by configuring it to "Load at startup".
2. Press shortkeys for quick changes of FOV.

11.4.1   Section [FOV] in config.ini file

You can configure the plugin with its dialog (Fig. 11.4) or edit `config.ini` file by yourself for changes of the settings for the Field of View plugin – just make it carefully!

ID	Type	Default	Description
fov_quick_0	float	0.5	Value of FOV for the shortcut Ctrl + Alt + 0
fov_quick_1	float	180	Value of FOV for the shortcut Ctrl + Alt + 1
fov_quick_2	float	90	Value of FOV for the shortcut Ctrl + Alt + 2
fov_quick_3	float	60	Value of FOV for the shortcut Ctrl + Alt + 3
fov_quick_4	float	45	Value of FOV for the shortcut Ctrl + Alt + 4
fov_quick_5	float	20	Value of FOV for the shortcut Ctrl + Alt + 5
fov_quick_6	float	10	Value of FOV for the shortcut Ctrl + Alt + 6
fov_quick_7	float	5	Value of FOV for the shortcut Ctrl + Alt + 7
fov_quick_8	float	2	Value of FOV for the shortcut Ctrl + Alt + 8
fov_quick_9	float	1	Value of FOV for the shortcut Ctrl + Alt + 9

[1]SkyChart / Cartes du Ciel planetarium: `http://www.ap-i.net/skychart/en/start`

11.5 Pointer Coordinates Plugin

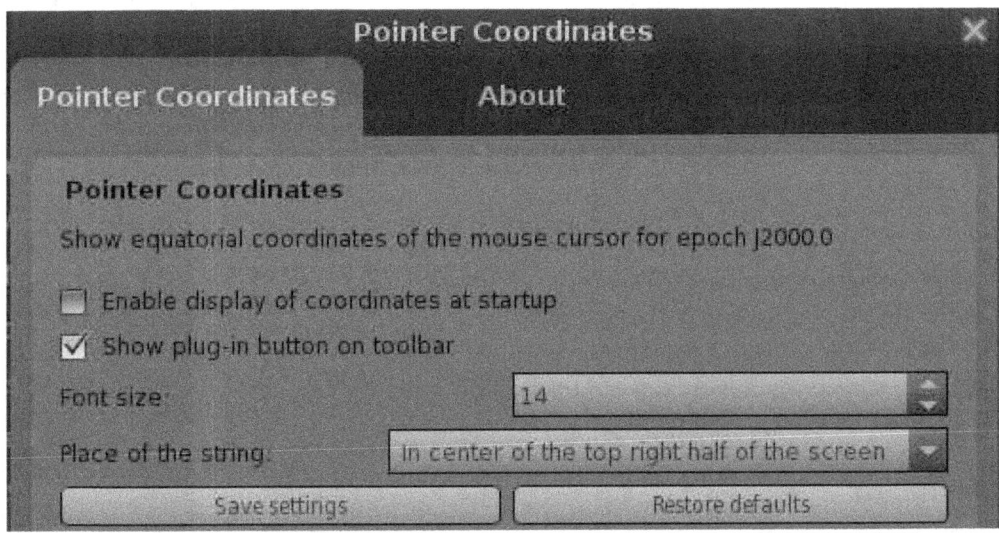

Figure 11.3: Interface of Pointer Coordinates plugin

The Pointer Coordinates plugin shows the coordinates of the mouse pointer. If enabled, click on the plugin button ☐ on the bottom toolbar to display the coordinates of the mouse pointer.

11.5.1 Section [PointerCoordinates] in config.ini file

You can edit config.ini file by yourself for changes of the settings for the Pointer Coordinates plugin – just make it carefully!

ID	Type	Description
enable_at_startup	bool	Enable displaying coordinates of the mouse pointer at startup of the plugin
flag_show_button	bool	Enable showing the button of the plugin on bottom toolbar
text_color	R,G,B	Color for text with coordinates of the mouse pointer
font_size	int	Font size for the displayed coordinates of the mouse pointer
current_displaying_place	string	Specifies the place of displaying coordinates of the mouse pointer. *Possible values*: TopRight, TopCenter, RightBottomCorner, Custom. *Default value*: TopRight.
current_coordinate_system	string	Specifies the coordinate system. *Possible values*: RaDecJ2000, RaDec, HourAngle, Ecliptic, AltAzi, Galactic. *Default value*: RaDecJ2000.
custom_coordinates	int,int	Specifies the screen coordinates of the custom place for displaying coordinates of the mouse pointer

11.6 Text User Interface

This plugin re-implements the "TUI" of the pre-0.10 versions of Stellarium, an unobtrusive menu used primarily by planetarium system operators to change settings, run scripts and so on.

Given that color configuration for lines and texts cannot be found elsewhere, an interesting part for general use (at least for extended configuration sessions) is the way to interactively select colors in this menu.

A full list of the commands for the TUI plugin is given in section 11.6.2.

11.6.1 Using the Text User Interface

1. Activate the text menu using the Alt + T key.[2]
2. Navigate the menu using the cursors keys.
3. To edit a value, press the right cursor until the value you wish to change it highlighted with > and < marks, e.g. >3.142<. Then press the cursor keys ↑ and ↓ to change the value. You may also type in a new value with the other keys on the keyboard.

11.6.2 TUI Commands

1	Location	(menu group)
1.1	Latitude	Set the latitude of the observer in degrees
1.2	Longitude	Set the longitude of the observer in degrees
1.3	Altitude	Set the altitude of the observer in meters
1.4	Solar System Body	Select the solar system body on which the observer is
2	Set Time	(menu group)
2.1	Current date/time	Set the time and date for which Stellarium will generate the view
2.2	Set Time Zone	(disabled in 0.15)
2.3	Days keys	(disabled in 0.15)
2.4	Startup date/time preset	Select the time which Stellarium starts with (if the "Sky Time At Start-up" setting is "Preset Time"
2.5	Startup date and time	The setting "system" sets Stellarium's time to the computer clock when Stellarium runs. The setting "preset" selects a time set in menu item "2.4 - Startup date/time preset"
2.6	Date Display Format	Change how Stellarium formats date values. "system_default" takes the format from the computer settings, or it is possible to select "yyyymmdd", "ddmmyyyy" or "mmddyyyy" modes

---

[2]This used to be hard-coded to M before version 0.15, but Alt + T is better to remember as it runs parallel with Ctrl + T for switching the GUI panels, and frees up M for the Milky Way. The Alt + T keybinding is hardcoded, i.e., cannot be reconfigured by the user, and should not be used for another function.

2.7	Time Display Format	Change how Stellarium formats time values. "system_default" takes the format from the computer settings, or it is possible to select "24h" or "12h" clock modes
3	General	(menu group)
3.1	Starlore	Select the sky culture to use (changes constellation lines, names, artwork)
3.2	Sky Language	Change the language used to describe objects in the sky
3.3	App Language	Change the application language (used in GUIs)
4	Stars	(menu group)
4.1	Show stars	Turn on/off star rendering
4.2	Relative Scale	Change the relative brightness of the stars. Larger values make bright stars much larger.
4.3	Absolute Scale	Change the absolute brightness of the stars. Large values show more stars. Leave at 1 for realistic views.
4.4	Twinkle	Sets how strong the star twinkling effect is - zero is off, the higher the value the more the stars will twinkle.
5	Colors	(menu group)
5.1	Constellation lines	Changes the colour of the constellation lines
5.2	Constellation labels	Changes the colour of the labels used to name stars
5.3	Art brightness	Changes the brightness of the constellation art
5.4	Constellation boundaries	Changes the colour of the constellation boundary lines
5.5	Cardinal points	Changes the colour of the cardinal points markers
5.6	Planet labels	Changes the colour of the labels for planets
5.7	Planet orbits	Changes the colour of the orbital guide lines for planets
5.8	Planet trails	Changes the colour of the planet trails lines
5.9	Meridian Line	Changes the colour of the meridian line
5.10	Azimuthal Grid	Changes the colour of the lines and labels for the azimuthal grid
5.11	Equatorial Grid	Changes the colour of the lines and labels for the equatorial grid
5.12	Equatorial J2000 Grid	Changes the colour of the lines and labels for the equatorial J2000.0 grid

5.13	Equator Line	Changes the colour of the equator line
5.14	Ecliptic Line	Changes the colour of the ecliptic line
5.15	Ecliptic Line (J2000)	Changes the colour of the J2000 ecliptic line
5.16	Nebula names	Changes the colour of the labels for nebulae
5.17	Nebula hints	Changes the colour of the circles used to denote the positions of unspecified nebulae
5.18	Galaxy hints	Changes the colour of the ellipses used to denote the positions of galaxies
5.19	Bright nebula hints	Changes the colour of the squares used to denote the positions of bright nebulae
5.20	Dark nebula hints	Changes the colour of the squares used to denote the positions of dark nebulae
5.21	Clusters hints	Changes the colour of the symbols used to denote the positions of clusters
5.22	Horizon line	Changes the colour of the horizon line
5.23	Galactic grid	Changes the colour of the galactic grid
5.24	Galactic equator line	Changes the colour of the galactic equator line
5.25	Opposition/conjunction longitude line	Changes the colour of the opposition/conjunction line
6	Effects	(menu group)
6.1	Light Pollution	Changes the intensity of the light pollution (see Appendix B Bortle Scale index)
6.2	Landscape	Select the landscape which Stellarium draws when ground drawing is enabled. Press ⏎ to activate.
6.3	Setting Landscape Sets Location	If "Yes" then changing the landscape will move the observer location to the location for that landscape (if one is known). Setting this to "No" means the observer location is not modified when the landscape is changed.
6.4	Auto zoom out returns to initial … view	Changes the behaviour when zooming out from a selected object. When set to "Off", selected object will stay in center. When set to "On", view will return to startup view.
6.5	Zoom Duration	Sets the time for zoom operations to take (in seconds)
6.6	Milky Way intensity	Changes the brightness of the Milky Way
6.7	Zodiacal light intensity	Changes the brightness of the Zodiacal light
7	Scripts	(menu group)

7.1	Run local script	Run a script from the scripts sub-directory of the User Directory or Installation Directory (see section 5 (Files and Directories))
7.1	Stop running script	Stop execution of a currently running script
8	Administration	(menu group)
8.1	Load default configuration	Reset all settings according to the main configuration file
8.2	Save current configuration	Save the current settings to the main configuration file
8.3	Shutdown	Emits a command configured in

11.6.3 Section [tui] in config.ini file

The section in config.ini for this plugin is named only [tui] for historical reasons. As always, be careful when editing!

ID	Type	Description
tui_font_color	R,G,B	Font color for TUI text
tui_font_size	int	Font size for the TUI
flag_show_gravity_ui	bool	Bend menu text around the screen center. May be useful in planetarium setups, and should then be used together with "Disc viewport" in the configuration menu (see 4.3.4).
flag_show_tui_datetime	bool	Show date and time in lower center.
flag_show_tui_short_obj_info	bool	Show some object info in lower right, or (in planetarium setups with "Disc viewport" active,) wrapped along the outer circle border.
admin_shutdown_cmd	string	executable command to shutdown your system. Best used on Linux or Mac systems. E.g. shutdown -h now

11.7   Remote Control

The Remote Control plugin was developed in 2015 during the ESA Summer of Code in Space initiative. It enables the user to control Stellarium through an external web interface using a standard web browser like Firefox or Chrome, instead of using the main GUI. This works on the same computer Stellarium runs as well as over the network. Even more, multiple "remote controls" can access the same Stellarium instance at the same time, without getting in the way of each other. Much of the functionality the main interface provides is already available through it, and it is still getting extended.

The plugin may be useful for presentation scenarios, hiding the GUI from the audience and allowing the presenter to change settings on a separate monitor without showing distracting dialog windows. It also allows to start and stop scripts remotely. Because the web interface can be customized (or completely replaced) with some knowledge of HTML, CSS and JavaScript, another possibility is a kiosk mode, where untrusted users can execute a variety of predefined actions (like starting recorded tours) without having access to all Stellarium settings. The web API can also be accessed directly (without using a browser and the HTML interface), allowing control of Stellarium with external programs and scripts using HTTP calls like with the tools `wget` and `curl`.

11.7.1   Using the plugin

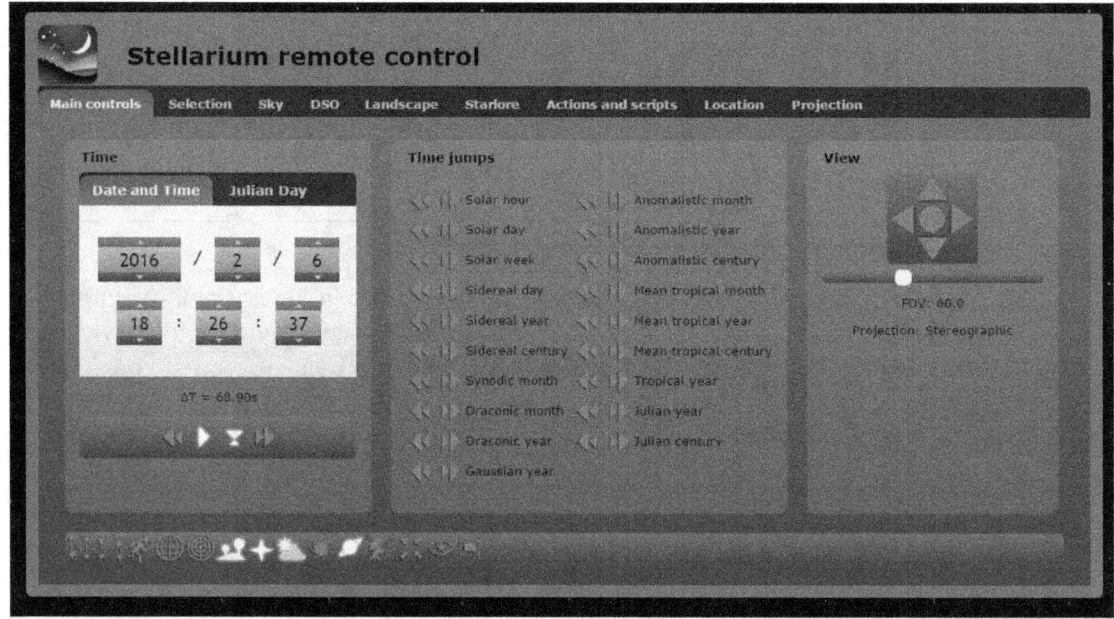

Figure 11.4: The default remote control web interface

After enabling the plugin, you can set it up through the configuration dialog. When "Enable automatically on startup" is checked (it is by default), the web server is automatically started whenever Stellarium starts. You can also manually start/stop the server using the "Server enabled" checkbox and the button 🖳 in the toolbar.

The plugin starts a HTTP server on the specified port. The default port is 8090, so you can try to reach the remote control after enabling it by starting a browser on the same computer and entering `http://localhost:8090` in the address bar. When trying to access the remote control from another computer, you need the IP address or the hostname of the server on which Stellarium runs. The plugin shows the locally detected address, but depending on your network or if you need external access you might need to use a different one — contact your network administrator if you need help with that.

The access to the remote control may optionally be restricted with a simple password.

Warning: *currently no network encryption is used, meaning that an attacker having access to your network can easily find out the password by waiting for a user entering it. Access from the Internet to the plugin should generally be restricted, except if countermeasures such as VPN usage are taken! If you are in a home network using NAT (network access translation), this should be enough for basic security except if port forwarding or a DMZ is configured.*

11.7.2 Remote Control Web Interface

If you are familiar with the main Stellarium interface, you should easily find your way around the web interface. Tabs at the top allow access to different settings and controls. The remote control automatically uses the same language as set in the main program.

The contents of the various tabs:

Main Contains the time controls and most of the buttons of the main bottom toolbar. An additional control allows moving the view like when dragging the mouse or using the arrow keys in Stellarium, and a slider enables the changing of the field of view. There are also buttons to quickly execute time jumps using the commonly used astronomical time intervals.

Selection Allows searching and selecting objects like in section 4.5. SIMBAD search is also supported. Quick select buttons are available for the primary solar system objects. It also displays the information text for current selection.

Sky Settings related to the sky display as shown in the "View" dialog as shown in subsection 4.4.1.

DSO The deep-sky object catalog, filter and display settings like in subsection 4.4.3.

Landscape Changing and configuring the background landscape, see subsection 4.4.4

Actions and scripts Lists all registered actions, and allows starting and stopping of scripts (chapter 15). If there is no button for the action you want in another tab, you can find all actions which can be configured as a keyboard shortcut (section 4.3) here.

Location Allows changing the location, like in section 4.2. Custom location saving is currently not supported.

Projection Switch the projection method used, like subsection 4.4.3.

11.7.3 Remote Control Commandline API

It is also possible to send commands via command line, e.g.,

```
wget -q --post-data 'id=show.ssc' http://stella:8090/api/scripts/run >/dev/null 2>&1
curl --data 'id=myScript.ssc' http://localhost:8090/api/scripts/run >/dev/null 2>&1
curl -d    'id=myScript.ssc' http://localhost:8090/api/scripts/run >/dev/null 2>&1
```

This allows triggering automatic show setups for museums etc. via some centralized schedulers like cron.

11.7.4 Developer information

If you are a developer and would like to add functionality to the Remote Control API, customize the web interface or access the API through another program, further information can be found in the plugin's developer documentation.

11.8 Solar System Editor Plugin

Stellarium stores its data (orbital elements and other details) about solar system objects (planets, their moons, minor bodies) in the file data/ssystem.ini. The file will be taken from the user data directory if it also exists there, which means, users can add minor planets or comets as they become observable by editing this file.

This plugin provides a window to the Minor Planet Center (MPC) where the latest Solar System information can be found. When this plugin is loaded (see section 10.1) the first tab allows to import, export or reset your ssystem.ini.

The second tab lists all currently loaded objects. It is recommended to remove old entries of last year's comets if you don't need them any longer. If you have a very weak computer, you may want to reduce the number of minor bodies (and maybe even planetary moons) to improve performance.

On this tab, you find the option to connect to the MPC and download current orbital elements.

11.9 Timezone Configuration Plugin

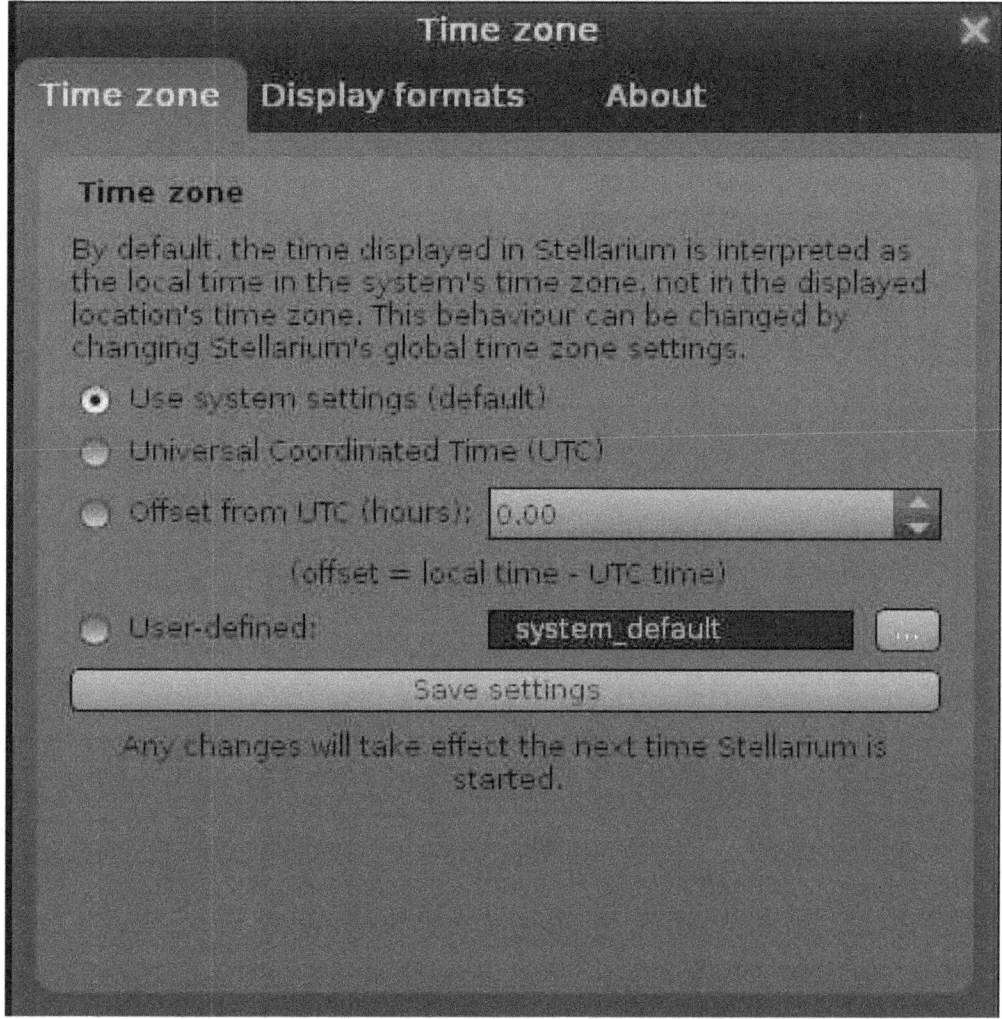

Figure 11.5: Interface of the TimeZone Configuration Plugin

After installation, Stellarium uses the timezone configured in the computer's operating system. This may have unwanted consequences, e.g. when you want to check the sky for a location on another continent. You may set location in the location panel (see 4.2), but the time will not be the zone time of this location but still zone time of your computer!

Another effect of our civilisation that sometimes brings unwanted side effects concerns daylight saving time (DST): When you run a timelapse with 1-day intervals, you will see what appear to be wrong jumps on the days when DST is switched (e.g., end of March/end of October, but rules evolve over time and regions). You may want to keep a single time zone without DST for such simulations.

If enabled (see section 10.1), you can choose to use your timezone (default), or plain Universal Coordinated Time (UTC) (equivalent to what used to be called Greenwich Mean Time GMT), or select a time zone (in steps of 15 minutes).

Note that you must exit and restart Stellarium to activate the new timezone, so jumping from location to location on Earth and showing correct zone time is not possible.

12. Object Catalog Plugins

Several plugins provide users with some more object classes.

12.1 Bright Novae Plugin

Figure 12.1: Nova Cygni 1975 (also known as **V1500 Cyg**)

The Bright Novae plugin provides visualization of some bright novae in the Milky Way galaxy. If enabled (see section 10.1), bright novae from the past will be presented in the sky at the correct times. For example, set date and time to 30 August 1975, look at the constellation *Cygnus* to see *Nova Cygni 1975*[1] (Fig. 12.1).

12.1.1 Section [Novae] in config.ini file

You can edit config.ini file by yourself for changes of the settings for the Bright Novae plugin – just make it carefully!

---

[1] http://en.wikipedia.org/wiki/V1500_Cygni

ID	Type	Description
last_update	string	Date and time of last update
update_frequency_days	int	Frequency of updates, in days
updates_enable	bool	Enable updates of bright novae catalog from Internet
url	string	URL of bright novae catalog

12.1.2   Format of bright novae catalog

To add a new nova, open a new line after line 5 and paste the following, note commas and brackets, they are important:

```
"Nova designation":
{
    "name": "name of nova",
    "type": "type of nova",
    "maxMagnitude": value of maximal visual magnitude,
    "minMagnitude": value of minimal visual magnitude,
    "peakJD": JD for maximal visual magnitude,
    "m2": Time to decline by 2mag from maximum (in days),
    "m3": Time to decline by 3mag from maximum (in days),
    "m6": Time to decline by 6mag from maximum (in days),
    "m9": Time to decline by 9mag from maximum (in days),
    "distance": value of distance between nova and
                Earth (in thousands of Light Years),
    "RA": "Right ascension (J2000)",
    "Dec": "Declination (J2000)"
},
```

For example, the record for **Nova Cygni 1975 (V1500 Cyg)** looks like:

```
"V1500 Cyg":
{
    "name": "Nova Cygni 1975",
    "type": "NA",
    "maxMagnitude": 1.69,
    "minMagnitude": 21,
    "peakJD": 2442655,
    "m2": 2,
    "m3": 4,
    "m6": 32,
    "m9": 263
    "distance": 6.36,
    "RA": "21h11m36.6s",
    "Dec": "48d09m02s"
},
```

12.1.3 Light curves

This plugin uses a very simple model for calculation of light curves for novae stars. This model is based on time for decay by N magnitudes from the maximum value, where N is 2, 3, 6 and 9. If a nova has no values for decay of magnitude then this plugin will use generalized values for it.

12.2 Historical Supernovae Plugin

Figure 12.2: Supernova 1604 (also known as **Kepler's Supernova**, **Kepler's Nova** or **Kepler's Star**)

Similar to the Historical Novae plugin (section 12.1), the Historical Supernovae plugin provides visualization of bright historical supernovae (Fig. 12.2) from the table below. If enabled (see section 10.1), bright supernovae from the past will be presented in the sky at the correct times. For example, set date and time to 29 April 1006, and look at the constellation *Lupus* to see *SN 1006A*.

12.2.1 List of supernovae in default catalog

Supernova	Date of max. brightness	Max. apparent mag.	Type	Name
SN 185A[2]	7 December	-6.0	Ia	
SN 386A	24 April	1.5	II	
SN 1006A[3]	29 April	-7.5	I	
SN 1054A[4]	3 July	-6.0	II	
SN 1181A[5]	4 August	-2.0	II	
SN 1572A[6]	5 November	-4.0	I	Tycho's Supernova
SN 1604A[7]	8 October	-2.0	I	Kepler's Supernova
SN 1680A[8]	15 August	6.0	IIb	Cassiopeia A
SN 1885A[9]	17 August	5.8	IPec	S Andromedae
SN 1895B	5 July	8.0	I	
SN 1920A	17 December	11.7	II	
SN 1921C	11 December	11.0	I	
SN 1937C	21 August	8.5	Ia	

SN 1960F	21 April	11.6	Ia	
SN 1960R	19 December	12.0	I	
SN 1961H	8 May	11.8	Ia	
SN 1962M	26 November	11.5	II	
SN 1966J	2 December	11.3	I	
SN 1968L	12 July	11.9	IIP	
SN 1970G	30 July	11.4	IIL	
SN 1971I	29 May	11.9	Ia	
SN 1972E[10]	8 May	8.4	Ia	
SN 1979C	15 April	11.6	IIL	
SN 1980K	31 October	11.6	IIL	
SN 1981B	9 March	12.0	Ia	
SN 1983N	17 July	11.4	Ib	
SN 1987A[11]	24 February	2.9	IIPec	
SN 1989B	6 February	11.9	Ia	
SN 1991T	26 April	11.6	IaPec	
SN 1993J[12]	30 March	10.8	IIb	
SN 1994D	31 March	11.8	Ia	
SN 1998bu	21 May	11.9	Ia	
SN 2004dj	31 July	11.3	IIP	
SN 2011fe[13]	13 September	10.06	Ia	
SN 2013aa	13 February	11.9	Ia	

12.2.2 Light curves

In this plugin a simple model of light curves for different supernovae has been implemented. A typical light curve used in the plugin for supernova type I is shown in Fig. 12.3 (bottom scale in days).

For supernova type II we use a typical light curve with plateau, which you can see in Fig. 12.4

---

[2]https://en.wikipedia.org/wiki/SN_185
[3]https://en.wikipedia.org/wiki/SN_1006
[4]https://en.wikipedia.org/wiki/SN_1054
[5]https://en.wikipedia.org/wiki/SN_1181
[6]https://en.wikipedia.org/wiki/SN_1572
[7]https://en.wikipedia.org/wiki/SN_1604
[8]https://en.wikipedia.org/wiki/Cassiopeia_A
[9]https://en.wikipedia.org/wiki/S_Andromedae
[10]https://en.wikipedia.org/wiki/SN1972e
[11]https://en.wikipedia.org/wiki/SN_1987A
[12]https://en.wikipedia.org/wiki/SN_1993J
[13]https://en.wikipedia.org/wiki/SN_2011fe

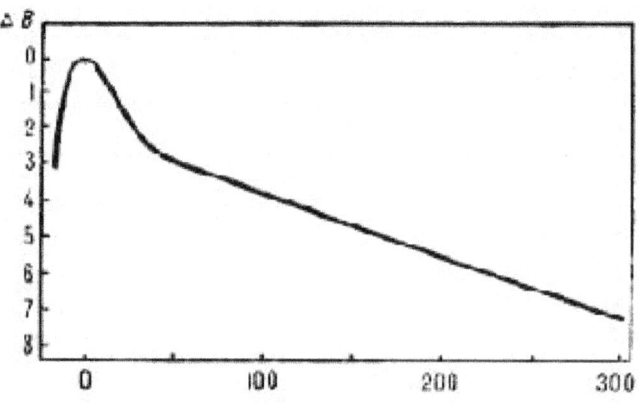

Figure 12.3: Light Curve of Supernova Type I

(bottom scale in days).

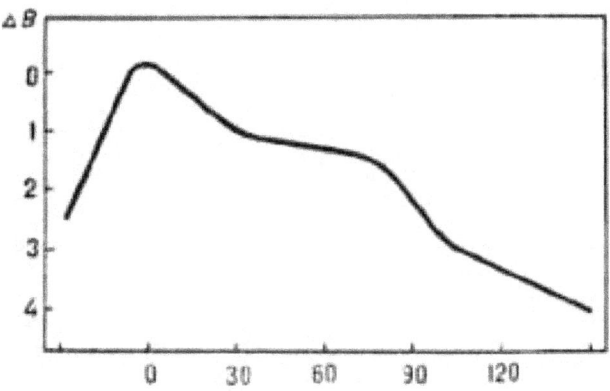

Figure 12.4: Light Curve of Supernova Type II

In both images for light curves the maximum brightness is marked as day 0.

12.2.3 Section [Supernovae] in config.ini file

You can edit config.ini file by yourself for changes of the settings for the Historical Supernovae plugin – just make it carefully!

ID	Type	Description
last_update	string	Date and time of last update
update_frequency_days	int	Frequency of updates, in days
updates_enable	bool	Enable updates of bright novae catalog from Internet
url	string	URL of bright novae catalog

12.2.4 Format of historical supernovae catalog

To add a new nova, open a new line after line 5 and paste the following, note commas and brackets, they are important:

```
"Supernova␣designation":
{
    "type": "type␣of␣supernova",
    "maxMagnitude": value of maximal visual magnitude,
    "peakJD": JD for maximal visual magnitude,
    "alpha": "Right␣ascension␣(J2000)",
    "delta": "Declination␣(J2000)",
    "distance": value of distance between supernova and
                Earth (in thousands of Light Years),
    "note": "notes␣for␣supernova"
},
```

For example, the record for **SN 1604A** (**Kepler's Supernova**) looks like:

```
"1604A":
{
    "type": "I",
    "maxMagnitude": -2,
    "peakJD": 2307190,
    "alpha": "17h30m36.00s",
    "delta": "-21d29m00.0s",
    "distance": 14,
    "note": "Kepler's␣Supernova"
},
```

12.3 Exoplanets Plugin

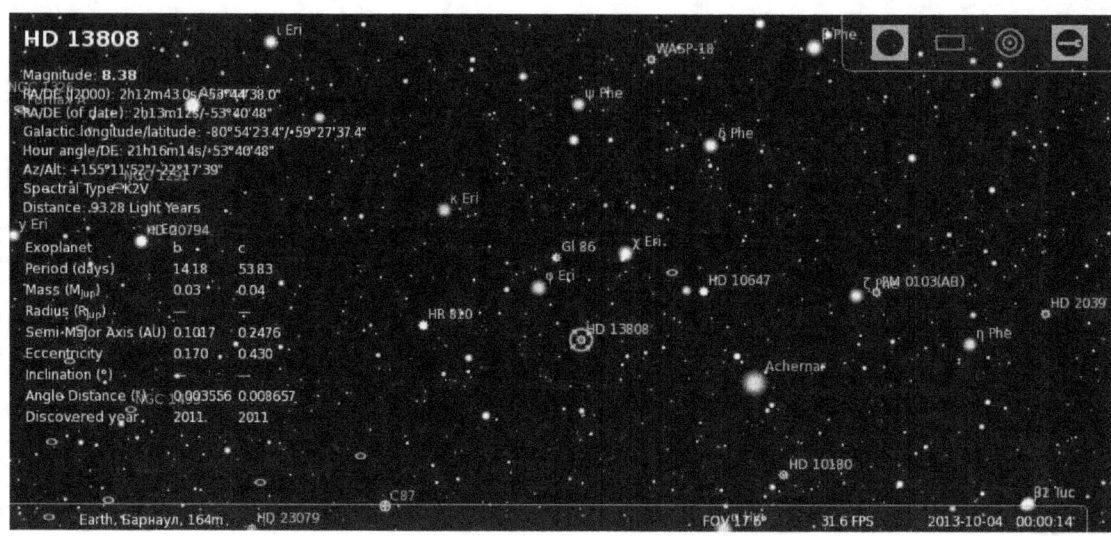

Figure 12.5: Planetary system HD 13808

This plugin plots the position of stars with exoplanets. Exoplanets data is derived from "The Extrasolar Planets Encyclopaedia"[14]. List of potential habitable exoplanets and data about them were taken from "The Habitable Exoplanets Catalog"[15] by the Planetary Habitability Laboratory[16].

If enabled (see section 10.1), just click on the Exoplanet button ⬚ on the bottom toolbar to display markers for the stars with known exoplanets. You can then either click on such a marked star or find the stars with exoplanets by their designation (e.g., *24 Sex*) in the F3 dialog (see 4.5).

12.3.1 Potential habitable exoplanets

This plugin can display potential habitable exoplanets (orange marker) and some information about those planets.

Planetary Class Planet classification from host star spectral type (F, G, K, M; see section 17.2.4), habitable zone (hot, warm, cold) and size (miniterran, subterran, terran, superterran, jovian, neptunian) (Earth = G-Warm Terran).

Equilibrium Temperature The planetary equilibrium temperature[17] is a theoretical temperature (in °C) that the planet would be at when considered simply as if it were a black body being heated only by its parent star (assuming a 0.3 bond albedo). As example the planetary equilibrium temperature of Earth is -18.15°C (255 K).

Earth Similarity Index (ESI) Similarity to Earth[18] on a scale from 0 to 1, with 1 being the most Earth-like. ESI depends on the planet's radius, density, escape velocity, and surface temperature.

---

[14]http://exoplanet.eu/
[15]http://phl.upr.edu/projects/habitable-exoplanets-catalog
[16]http://phl.upr.edu/home
[17]http://lasp.colorado.edu/~bagenal/3720/CLASS6/6EquilibriumTemp.html
[18]http://phl.upr.edu/projects/earth-similarity-index-esi

12.3.2 Proper names

In December 2015, the International Astronomical Union (IAU) has officially approved names for several exoplanets after a public vote.

Veritate * (14 And) – From the latin Veritas, truth. The ablative form means *where there is truth*[19].

Spe * (14 And b) – From the latin Spes, hope. The ablative form means *where there is hope*.

Musica (18 Del) – Musica is Latin for *music*.

Arion (18 Del b) – Arion was a genius of poetry and music in ancient Greece. According to legend, his life was saved at sea by dolphins after attracting their attention by the playing of his kithara.

Fafnir (42 Dra) – Fafnir was a Norse mythological dwarf who turned into a dragon.

Orbitar (42 Dra b) – Orbitar is a contrived word paying homage to the space launch and orbital operations of NASA.

Chalawan (47 UMa) – Chalawan is a mythological crocodile king from a Thai folktale.

Taphao Thong (47 UMa b) – Taphao Thong is one of two sisters associated with the Thai folk tale of Chalawan.

Taphao Kaew (47 UMa c) – Taphao Kae is one of two sisters associated with the Thai folk tale of Chalawan.

Helvetios (51 Peg) – Helvetios is Celtic for the *Helvetian* and refers to the Celtic tribe that lived in Switzerland during antiquity.

Dimidium (51 Peg b) – Dimidium is Latin for *half*, referring to the planet's mass of at least half the mass of Jupiter.

Copernicus (55 Cnc) – Nicolaus Copernicus or Mikolaj Kopernik (1473-1543) was a Polish astronomer who proposed the heliocentric model of the solar system in his book "De revolutionibus orbium coelestium".

Galileo (55 Cnc b) – Galileo Galilei (1564-1642) was an Italian astronomer and physicist often called the *father of observational astronomy* and the *father of modern physics*. Using a telescope, he discovered the four largest satellites of Jupiter, and the reported the first telescopic observations of the phases of Venus, among other discoveries.

Brahe (55 Cnc c) – Tycho Brahe (1546-1601) was a Danish astronomer and nobleman who recorded accurate astronomical observations of the stars and planets. These observations were critical to Kepler's formulation of his three laws of planetary motion.

Lipperhey * (55 Cnc d) – Hans Lipperhey (1570-1619) was a German-Dutch lens grinder and spectacle maker who is often attributed with the invention of the refracting telescope in 1608[20].

Janssen (55 Cnc e) – Jacharias Janssen (1580s-1630s) was a Dutch spectacle maker who is often attributed with invention of the microscope, and more controversially with the invention of the telescope.

Harriot (55 Cnc f) – Thomas Harriot (ca. 1560-1621) was an English astronomer, mathematician, ethnographer, and translator, who is attributed with the first drawing of the Moon through telescopic observations.

Amateru * (ε Tau b) – *Amateru* is a common Japanese appellation for shrines when they enshrine Amaterasu, the Shinto goddess of the Sun, born from the left eye of the god Izanagi[21].

Hypatia (ι Dra b) – Hypatia was a famous Greek astronomer, mathematician, and philosopher. She was head of the Neo-Platonic school at Alexandria in the early 5th century, until murdered

---

[19]The original name proposed, Veritas, is that of an asteroid important for the study of the solar system.

[20]The original spelling of Lippershey was corrected to Lipperhey on 15.01.2016. The commonly seen spelling Lippershey (with an s) results in fact from a typographical error dating back from 1831, thus should be avoided.

[21]The name originally proposed, Amaterasu, is already used for an asteroid.

by a Christian mob in 415.

Ran * (ε Eri) – Ran is the Norse goddess of the sea, who stirs up the waves and captures sailors with her net.

AEgir * (ε Eri b) – AEgir is Ran's husband, the personified god of the ocean. *AEgir* and *Ran* both represent the *Jotuns* who reign in the outer Universe; together they had nine daughters[22].

Tadmor * (γ Cep b) – Ancient Semitic name and modern Arabic name for the city of Palmyra, a UNESCO World Heritage Site.

Dagon (α PsA b) – Dagon was a Semitic deity, often represented as half-man, half-fish.

Tonatiuh (HD 104985) – Tonatiuh was the Aztec god of the Sun.

Meztli (HD 104985 b) – Meztli was the Aztec goddess of the Moon.

Ogma * (HD 149026) – Ogma was a deity of eloquence, writing, and great physical strength in the Celtic mythologies of Ireland and Scotland, and may be related to the Gallo-Roman deity *Ogmios*[23].

Smertrios (HD 149026 b) – Smertrios was a Gallic deity of war.

Intercrus (HD 81688) – Intercrus means *between the legs* in Latin style, referring to the star's position in the constellation Ursa Major.

Arkas (HD 81688 b) – Arkas was the son of Callisto (Ursa Major) in Greek mythology.

Cervantes (μ Ara) – Miguel de Cervantes Saavedra (1547-1616) was a famous Spanish writer and author of "El Ingenioso Hidalgo Don Quixote de la Mancha".

Quijote (μ Ara b) – Lead fictional character from Cervantes's "El Ingenioso Hidalgo Don Quixote de la Mancha".

Dulcinea (μ Ara c) — Fictional character and love interest of Don Quijote (or Quixote) in Cervantes's "El Ingenioso Hidalgo Don Quixote de la Mancha".

Rocinante (μ Ara d) – Fictional horse of Don Quijote in Cervantes's "El Ingenioso Hidalgo Don Quixote de la Mancha".

Sancho (μ Ara e) – Fictional squire of Don Quijote in Cervantes's "El Ingenioso Hidalgo Don Quixote de la Mancha".

Thestias * (β Gem b) – Thestias is the patronym of Leda and her sister Althaea, the daughters of Thestius. Leda was a Greek queen, mother of Pollux and of his twin Castor, and of Helen and Clytemnestra[24].

Lich (PSR B1257+12) – Lich is a fictional undead creature known for controlling other undead creatures with magic.

Draugr (PSR B1257+12 b) – Draugr refers to undead creatures in Norse mythology.

Poltergeist (PSR B1257+12 c) – Poltergeist is a name for supernatural beings that create physical disturbances, from German for noisy ghost.

Phobetor (PSR B1257+12 d) – Phobetor is a Greek mythological deity of nightmares, the son of Nyx, the primordial deity of night.

Titawin (υ And) – Titawin (also known as Medina of Tetouan) is a settlement in northern Morocco and UNESCO World Heritage Site. Historically it was an important point of contact between two civilizations (Spanish and Arab) and two continents (Europe and Africa) after the 8th century.

Saffar (υ And b) – Saffar is named for Abu al-Qasim Ahmed Ibn-Abd Allah Ibn-Omar al Ghafiqi Ibn-al-Saffar, who taught arithmetic, geometry, and astronomy in 11th century Cordova in Andalusia (modern Spain), and wrote an influential treatise on the uses of the astrolabe.

---

[22]Note the typographical difference between AEgir and Aegir, the Norwegian transliteration. The same name, with the spelling Aegir, has been attributed to one of Saturn's satellites, discovered in 2004.

[23]Ogmios is a name already attributed to an asteroid.

[24]The original proposed name Leda is already attributed to an asteroid and to one of Jupiter's satellites. The name Althaea is also attributed to an asteroid.

Samh (υ And c) – Samh is named for Abu al-Qasim 'Asbagh ibn Muhammad ibn al-Samh al-Mahri (or Ibn al-Samh), a noted 11th century astronomer and mathematician in the school of al Majriti in Cordova (Andalusia, now modern Spain).

Majriti (υ And d) – Majriti is named for Abu al-Qasim al-Qurtubi al-Majriti, a notable mathematician, astronomer, scholar, and teacher in 10th century and early 11th century Andalusia (modern Spain).

Libertas * (ξ Aql) – Libertas is Latin for liberty. Liberty refers to social and political freedoms, and a reminder that there are people deprived of liberty in the world even today. The constellation Aquila represents an eagle – a popular symbol of liberty.

Fortitudo * (ξ Aql b) – Fortitudo is Latin for fortitude. Fortitude means emotional and mental strength in the face of adversity, as embodied by the eagle (represented by the constellation Aquila).

All names with asterix mark (*) are modified based on the original proposals, to be consistent with the IAU rules.

12.3.3 Section [Exoplanets] in config.ini file

You can edit config.ini file by yourself for changes of the settings for the Exoplanets plugin – just make it carefully!

ID	Type	Description
last_update	string	Date and time of last update
update_frequency_hours	int	Frequency of updates, in hours
updates_enable	bool	Enable updates of exoplanets catalog from Internet
url	string	URL of exoplanets catalog
flag_show_exoplanets_button	bool	Enable showing button of exoplanets on bottom bar
distribution_enabled	bool	Enable distribution mode of display
timeline_enabled	bool	Enable timeline mode of display
habitable_enabled	bool	Enable habitable mode of display
enable_at_startup	bool	Enable displaying exoplanets at startup of the plugin
exoplanet_marker_color	R,G,B	Color for marker of star with planetary system
habitable_exoplanet_marker_color	R,G,B	Color for marker of star with planetary system with potential habitable exoplanets

12.3.4 Format of exoplanets catalog

To add a new exoplanet system, open a new line after line 5 and paste the following, note commas and brackets, they are important:

```
"Star␣designation":
{
        "exoplanets":
        [
        {
                "mass": mass of exoplanet (M jup),
                "radius": radius of exoplanet (R jup),
                "period": period of exoplanet (days),
                "semiAxis": semi-major axis (AU),
                "eccentricity": orbit's eccentricity,
                "inclination": orbit's inclination (degree),
                "angleDistance": angle distance from star
                                (arcseconds),
                "discovered": exoplanet discovered year,
                "hclass": "habitable␣class",
                "MSTemp": mean surface temperature (K),
                "ESI": Earth Similarity Index (*100),
                "planetProperName": "proper␣name␣of␣planet",
                "planetName": "designation␣of␣planet"
        },
        {
                "mass": mass of exoplanet (M jup),
                "radius": radius of exoplanet (R jup),
                "period": period of exoplanet (days),
                "semiAxis": semi-major axis (AU),
                "eccentricity": orbit's eccentricity,
                "inclination": orbit's inclination (degree),
                "angleDistance": angle distance from star
                                (arcseconds),
                "discovered": exoplanet discovered year,
                "hclass": "habitable␣class",
                "MSTemp": mean surface temperature (K),
                "ESI": Earth Similarity Index (*100),
                "planetProperName": "proper␣name␣of␣planet",
                "planetName": "designation␣of␣planet"
        }
        ],
        "distance": value of distance to star (pc),
        "stype": "spectral␣type␣of␣star",
        "smass": value of mass of star (M sun),
        "smetal": value of metallicity of star,
        "Vmag": value of visual magnitude of star,
        "sradius": value of radius of star (R sun),
        "effectiveTemp": value of effective temperature
                        of star (K),
        "starProperName": "proper␣name␣of␣the␣star",
```

```
                "hasHP": boolean (has potential habitable planets),
                "RA": "Right␣ascension␣(J2000)",
                "DE": "Declination␣(J2000)"
},
```

For example, the record for *24 Sex* looks like:

```
"24␣Sex":
{
                "exoplanets":
                [
                {
                        "mass": 1.99,
                        "period": 452.8,
                        "semiAxis": 1.333,
                        "eccentricity": 0.09,
                        "angleDistance": 0.017821,
                        "discovered": 2010,
                        "planetName": "b"
                },
                {
                        "mass": 0.86,
                        "period": 883.0,
                        "semiAxis": 2.08,
                        "eccentricity": 0.29,
                        "angleDistance": 0.027807,
                        "discovered": 2010,
                        "planetName": "c"
                }
                ],
                "distance": 74.8,
                "stype": "G5",
                "smass": 1.54,
                "smetal": -0.03,
                "Vmag": 7.38,
                "sradius": 4.9,
                "effectiveTemp": 5098,
                "RA": "10h23m28s",
                "DE": "-00d54m08s"
},
```

12.4   Pulsars Plugin

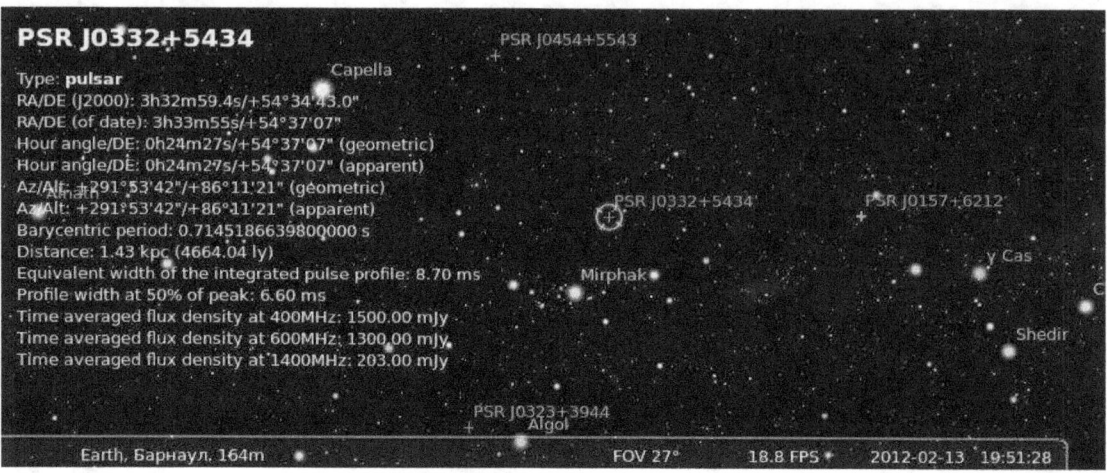

Figure 12.6: PSR J0332-5434

This plugin plots the position of various pulsars, with object information about each one. Pulsar data is derived from *The ATNF Pulsar Catalogue* [32].

If enabled (see section 10.1), use the `PSR` button to activate display of pulsars. The GUI allows a few configuration options. You can also find a pulsar (F3) by its designation (e.g., *PSR J0437-4715*).

12.4.1   Section [Pulsars] in config.ini file

ID	Type	Description
last_update	string	Date and time of last update
update_frequency_days	int	Frequency of updates, in days
updates_enable	bool	Enable updates of pulsars catalog from Internet
url	string	URL of pulsars catalog
enable_at_startup	bool	Enable displaying of pulsars at startup of Stellarium
distribution_enabled	bool	Enable distribution mode for the pulsars
flag_show_pulsars_button	bool	Enable displaying pulsars button on toolbar
marker_color	R,G,B	Color for marker of the pulsars
glitch_color	R,G,B	Color for marker of the pulsars with glitches
use_separate_colors	bool	Use separate colors for different types of the pulsars

12.4.2 Format of pulsars catalog

To add a new pulsar, open a new line after line 5 and paste the following, note commas and brackets, they are important:

```
"Pulsar␣designation":
{
    "RA": "Right␣ascension␣(J2000)",
    "DE": "Declination␣(J2000)",
    "notes": "type␣of␣pulsar",
    "distance": value of distance based on electron density
                model (kpc),
    "period": value of barycentric period of the pulsar (s),
    "parallax": value of annular parallax (mas),
    "bperiod": value of binary period of pulsar (days),
    "pderivative": value of time derivative of barcycentric
                period,
    "dmeasure": value of dispersion measure (cm^-3 pc),
    "frequency": value of barycentric rotation frequency (Hz),
    "pfrequency": value of time derivative of barycentric
                rotation frequency (s^-2)
    "eccentricity": value of eccentricity,
    "w50": value of profile width at 50% of peak (ms),
    "s400": value of time averaged flux density at
            400 MHz (mJy),
    "s600": value of time averaged flux density at
            600 MHz (mJy),
    "s1400": value of time averaged flux density at
            1400 MHz (mJy)
},
```

For example, the record for **PSR J0014+4746** looks like:

```
"PSR␣J0014+4746":
{
    "distance": 1.82,
    "dmeasure": 30.85,
    "frequency": 0.805997239145,
    "pfrequency": -3.6669E-16,
    "w50": 88.7,
    "s400": 14,
    "s600": 9,
    "s1400": 3,
    "RA": "00h14m17.75s",
    "DE": "47d46m33.4s"
},
```

12.5 Quasars Plugin

The Quasars plugin provides visualization of some quasars brighter than 16 visual magnitude. A catalogue of quasars compiled from *Quasars and Active Galactic Nuclei* (13th Ed.) [66].

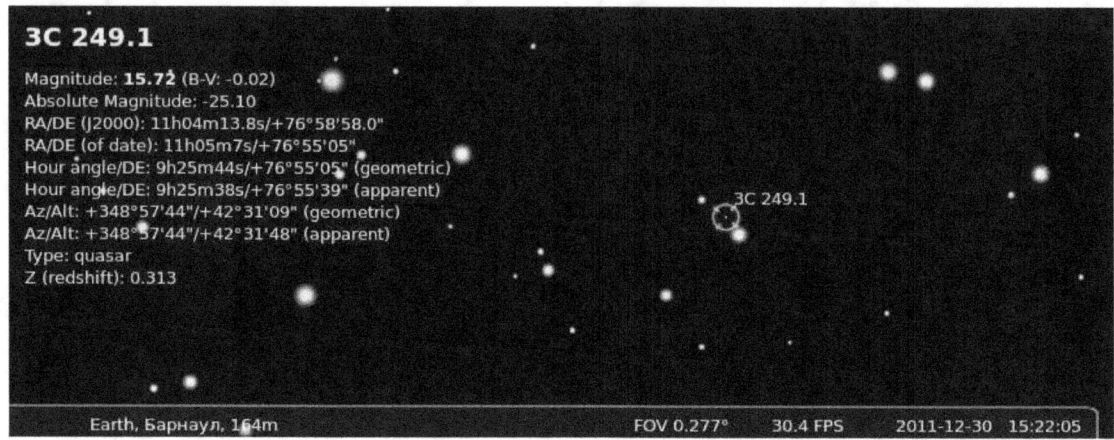

Figure 12.7: 3C 249.1, also known as LEDA 2821945 or 4C 77.09

If enabled (see section 10.1), use the QSO button to activate display of quasars. The GUI allows a few configuration options. You can also find a quasar (F3) by its designation (e.g., *3C 273*).

12.5.1 Section [Quasars] in config.ini file

ID	Type	Description
last_update	string	Date and time of last update
update_frequency_days	int	Frequency of updates, in days
updates_enable	bool	Enable updates of quasars catalog from Internet
url	string	URL of quasars catalog
enable_at_startup	bool	Enable displaying of quasars at startup of Stellarium
distribution_enabled	bool	Enable distribution mode for the quasars
flag_show_quasars_button	bool	Enable displaying quasars button on toolbar
marker_color	R,G,B	Color for marker of the quasars

12.5.2 Format of quasars catalog

To add a new quasar, open a new line after line 5 and paste the following, note commas and brackets, they are important:

```
"Quasar␣designation":
{
    "RA": "Right␣ascension␣(J2000)",
    "DE": "Declination␣(J2000)",
    "Amag": value of absolute magnitude,
    "Vmag": value of visual magnitude,
    "z": value of Z (redshift),
    "bV": value of B-V colour
},
```

For example, the record for **3C 249.1** looks like:

```
"3C␣249.1":
{
    "RA": "11h04m13.8s",
    "DE": "+76d58m58s",
    "Amag": -25.1,
    "Vmag": 15.72,
    "z": 0.313,
    "bV": -0.02
},
```

12.6   Meteor Showers Plugin

Figure 12.8: The 1833 Leonids replayed with the Meteor Showers plugin.

In contrast and extension of the random *shooting stars* feature of Stellarium (see section 17.6), this plugin provides data for real meteor showers and a marker for each active and inactive radiant, showing real information about its activity. If enabled (see section 10.1), just click on the Meteor Showers button ▱ on the bottom toolbar to display markers for the radiants.

12.6.1   Terms

Meteor shower

A *meteor shower* is a celestial event in which a number of meteors are observed to radiate, or originate, from one point in the night sky. These meteors are caused by streams of cosmic debris called *meteoroids* entering Earth's atmosphere at extremely high speeds on parallel trajectories. Most meteors are smaller than a grain of sand, so almost all of them disintegrate and never hit the Earth's surface. Intense or unusual meteor showers are known as meteor outbursts and meteor storms, which may produce greater than 1,000 meteors an hour.

Radiant

The *radiant* or *apparent radiant* of a meteor shower is the point in the sky from which (to a planetary observer) meteors appear to originate. The Perseids, for example, are meteors which appear to come from a point within the constellation of Perseus.

An observer might see such a meteor anywhere in the sky but the direction of motion, when traced back, will point to the radiant. A meteor that does not point back to the known radiant for a given shower is known as a sporadic and is not considered part of that shower.

Many showers have a radiant point that changes position during the interval when it appears. For example, the radiant point for the Delta Aurigids drifts by more than a degree per night.

Zenithal Hourly Rate (ZHR)

The *Zenithal Hourly Rate* (ZHR) of a meteor shower is the number of meteors a single observer would see in one hour under a clear, dark sky (limiting apparent magnitude of 6.5) if the radiant of the shower were at the zenith. The rate that can effectively be seen is nearly always lower and decreases the closer the radiant is to the horizon.

Population index

The *population index* indicates the magnitude distribution of the meteor showers. Values below 2.5 correspond to distributions where bright meteors are more frequent than average, while values above 3.0 mean that the share of faint meteors is larger than usual.

12.6.2 Section [MeteorShowers] in config.ini file

You can edit `config.ini` file by yourself for changes of the settings for the Meteor Showers plugin – just make it carefully!

ID	Type	Description
last_update	string	Date and time of last update
update_frequency_hours	int	Frequency of updates, in hours
updates_enable	bool	Enable updates of the meteor showers catalog from Internet
url	string	URL of the meteor showers catalog
flag_show_ms_button	bool	Enable showing button of the meteor showers on bottom bar
flag_show_radiants	bool	Enable displaying markers for the radiants of the meteor showers
flag_active_radiants	bool	Flag for displaying markers for the radiants of the active meteor showers only
enable_at_startup	bool	Enable displaying meteor showers at starup plugin
show_radiants_labels	bool	Flag for displaying labels near markers of the radiants of the meteor showers
font_size	int	Font size for label of markers of the radiants of the meteor showers
colorARG	R,G,B	Color for marker of active meteor showers with generic data
colorARR	R,G,B	Color for marker of active meteor showers with real data
colorIR	R,G,B	Color for marker of inactive meteor showers

12.6.3 **Format of Meteor Showers catalog**

To add a new meteor shower, you just need to:

1. Copy the code of some valid meteor shower;
2. Paste it in the line 6 (right after the "showers": {) of the showers.json document;
3. Replace the information according with your needs.

Note commas and brackets, they are very important! For example, below is a record for *Northern Taurids*:

```
"NTA":
          {
                  "designation": "Northern␣Taurids",
                  "activity":
                  [
                  {
                          "year": "generic",
                          "zhr": 5,
                          "start": "09.25",
                          "finish": "11.25",
                          "peak": "11.12"
                  },
                  {
                          "year": "2014",
                          "start": "10.20",
                          "finish": "12.10"
                  },
                  {
                          "year": "2013",
                          "start": "10.20",
                          "finish": "12.10"
                  },
                  {
                          "year": "2012",
                          "start": "10.20",
                          "finish": "12.10"
                  },
                  {
                          "year": "2011",
                          "start": "10.20",
                          "finish": "12.10"
                  }
                  ],
                  "speed": 29,
                  "radiantAlpha": "58",
                  "radiantDelta": "+22",
                  "driftAlpha": "5",
                  "driftDelta": "1",
                  "colors":
                  [
                  {
                          "color": "yellow",
```

```
                                "intensity": 80
                },
                {
                        "color": "white",
                        "intensity": 20
                }
                ],
                "parentObj": "Comet␣C/1917␣F1␣(Mellish)",
                "pidx": 2.3
        },
```

12.6.4 Further Information

You can get more info about meteor showers here:

- Wikipedia about Meteor showers: https://en.wikipedia.org/wiki/Meteor_Showers
- International Meteor Organization: http://www.imo.net/

Acknowledgements

This plugin was created as project of ESA Summer of Code in Space 2013[25].

---

[25]http://sophia.estec.esa.int/socis2013/?q=about

12.7   Navigational Stars Plugin

Figure 12.9: Navigational stars on the screen

This plugin marks navigational stars from a selected set:

Anglo-American — the 57 "selected stars" that are listed in *The Nautical Almanac*[26] jointly published by Her Majesty's Nautical Almanac Office and the US Naval Observatory since 1958; consequently, these stars are also used in navigational aids such as the *2102-D Star Finder*[27] and *Identifier*.

French — the 81 stars that are listed in the *Éphémérides Nautiques* published by the French Bureau des Longitudes.

Russian — the 160 stars that are listed in the Russian Nautical Almanac.

If enabled (see section 10.1), just click on the Sextant button ⚓ on the bottom toolbar to display markers for the navigational stars. This can help you in training your skills in astronomical navigation before you cruise the ocean in the traditional way, with your sextant and chronometer.

12.7.1   Section [NavigationalStars] in config.ini file

You can edit config.ini file by yourself for changes of the settings for the Navigational Stars plugin – just make it carefully!

ID	Type	Description
navstars_color	R,G,B	Color of markers of navigational stars
current_ns_set	string	Current set of navigational stars. Possible values: *AngloAmerican*, *French* and *Russian*.

---

[26]The Nautical Almanac website – http://aa.usno.navy.mil/publications/docs/na.php

[27]Rude Starfinder 2102-D description and usage instruction – http://oceannavigation.blogspot.ru/2008/12/rude-starfinder-2102-d.html

12.8 **Satellites Plugin**

The Satellites plugin displays the positions of artifical satellites in Earth's orbit based on a catalog of orbital data. It allows automatic updates from online sources and manages a list of update file URLs.

To calculate satellite positions, the plugin uses an implementation of the SGP4/SDP4 algorithms (J.L. Canales' **gsat** library), using as its input data in NORAD's two-line element set (TLE[28]) format. Lists with TLEs for hundreds of satellites are available online and are regularly updated. The plugin downloads the lists prepared by `http://celestrak.com` to keep itself up-to-date, but the users can specify other sources online or load updates from local files.

If enabled (see section 10.1), just click on the Satellite button on the bottom toolbar to display markers for the satellites.

It should now be possible to search for artificial satellites using the regular search dialog (F3). Note that at any given time, most Satellites will be below the horizon.

12.8.1 **Satellite Properties**

Name and identifiers Each satellite has a name. It's displayed as a label of the satellite hint and in the list of satellites. Names are not unique though, so they are used only for presentation purposes.

Satellite Catalog In the *Satellite Catalog* satellites are uniquely identified by their NORAD number, which is encoded in TLEs.

Grouping A satellite can belong to one or more groups such as "amateur", "geostationary" or "navigation". They have no other function but to help the user organize the satellite collection. Group names are arbitrary strings defined in the Satellite Catalog for each satellite and are more similar to the concept of tags than a hierarchical grouping. A satellite may also not belong to any group at all.

By convention, group names are in lowercase. The GUI translates some of the groups used in the default catalog.

12.8.2 **Satellite Catalog**

The satellite catalog is stored on the disk in JSON[29] format, in a file named `satellites.json`. A default copy is embedded in the plug-in at compile time. A working copy is kept in the user data directory.

To add a new satellite, open a new line after line 5 and paste the following, note commas and brackets, they are important:

```
"NORAD␣number":
{
  "name": "name␣of␣the␣satellite"
  "description": "description␣goes␣here",
  "comms": [
      {
          "description": "downlink␣1",
          "frequency": 437.49,
          "modulation": "AFSK␣1200␣bps"
      },
      {
          "description": "downlink␣2",
          "frequency": 145.825
      }
              ],
  "groups": ["group1", "group2"],
```

---

[28]TLE: `https://en.wikipedia.org/wiki/Two-line_element_set`
[29]`http://www.json.org/`

```
  "tle1": "1␣12345U␣90005D␣␣␣09080.85236265␣␣.00000014␣␣00000-0␣␣20602-4␣0␣␣5632",
  "tle2": "2␣12345␣98.2700␣␣53.2702␣0011918␣␣71.1776␣289.0705␣14.31818920␣␣␣653",
  "visible": true
},
```

Explanation of the fields:

NORAD number required parameter, surrounded by double quotes ("), followed by a colon (:). It is used internally to identify the satellite. You should replace the text "NORAD number" with the first number on both lines of the TLE set (in this case, "12345"). It must match the number of the satellite in the source you are adding from if you want the TLE to be automatically updated.

The remaining parameters should be listed between two curly brackets and the closing curly bracket must be followed by a comma to separate it from the next satellite in the list:

name required parameter. It will be displayed on the screen and used when searching for the satellite with the Find window. Use the description field for a more readable name if you like. (The description field can accept HTML tags such as
 (new line), <b></b> (bold), etc.)

description optional parameter, double quoted. Appears when you click on the satellite

comms optional parameter, square bracketed list of curly bracketed communications information.

groups optional parameter, comma separated list of double quoted group names contained in square brackets. Used for grouping satellites in the drop down box on the config (see above)

tle1 required, line 1 of the TLE, must be contained in double quotes and begin with "1 "

tle2 required, line 2 of the TLE, must be contained in double quotes and begin with "2 "

visible required parameter, set to true if you want to see it, this can be toggled from the configuration window once the satellite is loaded.

You can edit the tags for a satellite, modify the description and comms data, and even add new satellites.

12.8.3 Configuration

The plug-in's configuration data is stored in Stellarium's main configuration file.

12.8.4 Sources for TLE data

Celestrak [30] used as default update source, it also has TLE lists beyond those included by default in Satellite plug-in

TLE.info [31]

Space Track [32] the definitive source, requires signup, operated by United States Department of Defense

---

[30]http://celestrak.com/NORAD/elements/

[31]http://www.tle.info/joomla/index.php

[32]http://www.space-track.org/

12.9 ArchaeoLines Plugin

GEORG ZOTTI

Figure 12.10: Declination Lines provided by the ArchaeoLines plugin

12.9.1 Introduction

In the archaeoastronomical literature, several astronomically derived orientation schemes are prevalent. Often prehistorical and historical buildings are described as having been built with a main axis pointing to a sunrise on summer or winter solstice. There can hardly be a better tool than Scenery3D (see chapter 13) to investigate a 3D model of such a building, and this plugin has been introduced in version 0.13.3 as a further tool in the archaeoastronomer's toolbox[72].

When activated (see section 10.1), you can find a a tool bar button $\boxed{\text{🜋}}$ (in the shape of a *trilithon* with the sun shining through it). Press this, or $\boxed{\text{Ctrl}}$+$\boxed{\text{U}}$, to display the currently selected set of characteristical diurnal arcs.

12.9.2 Characteristic Declinations

The ArchaeoLines plugin displays any combination of declination arcs δ most relevant to archaeo- or ethnoastronomical studies. Of course, principles used in this context are derived from natural observations, and many of these declinations are still important in everyday astronomy.

- Declinations of equinoxes (i.e., the equator, $\delta = 0$) and the solstices ($\delta = \pm\varepsilon$)
- Declinations of the crossquarter days (days between solstices and equinoxes, $\delta = \pm\varepsilon/\sqrt{2}$)
- Declinations of the Major Lunar Standstills ($\delta = \pm(\varepsilon + i)$)
- Declinations of the Minor Lunar Standstills ($\delta = \pm(\varepsilon - i)$)
- Declination of the Zenith passage ($\delta = \varphi$)
- Declination of the Nadir passage ($\delta = -\varphi$)
- Declination δ of the currently selected object
- Current declination $\delta_\odot$ of the sun
- Current declination $\delta_\leftmoon$ of the moon
- Current declination δ_P of a naked-eye planet

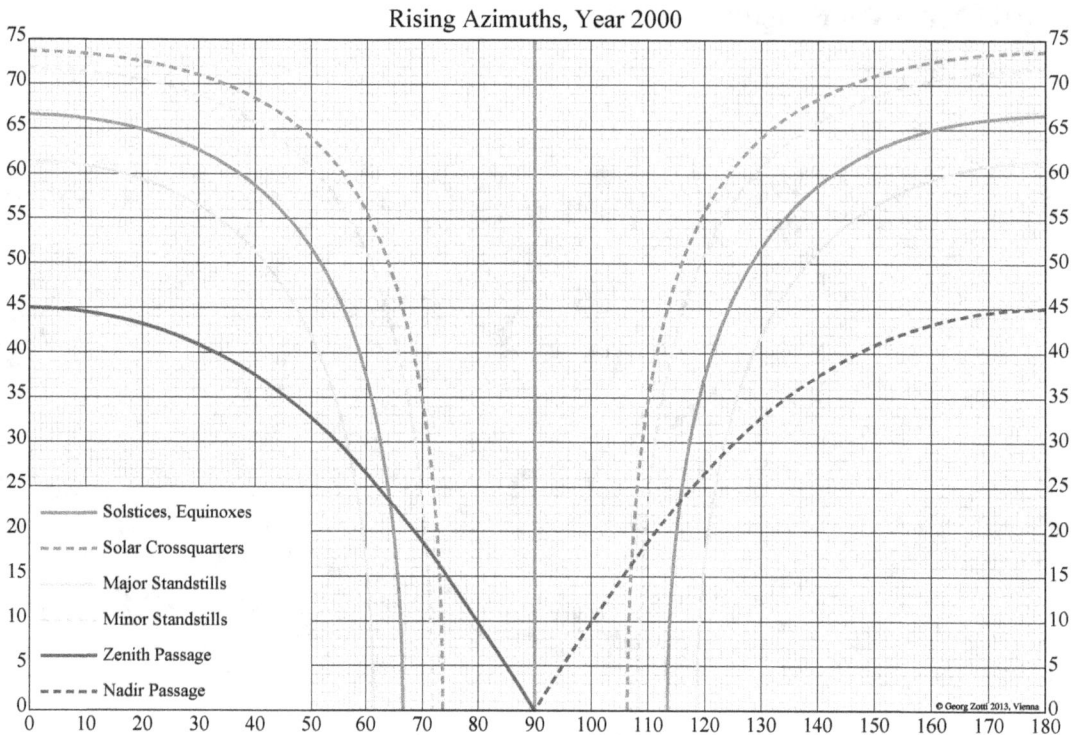

Figure 12.11: Rising azimuths of a few important events for sun and moon, and zenith and nadir passages depending on geographic latitudes (vertical axis).

The principal relation between declinations δ, geographic latitude φ, and the rising azimuth A is computed from

$$\cos A = -\frac{\sin \delta}{\cos \varphi}. \tag{12.1}$$

This formula does not take into account local horizon elevation nor atmospheric refraction nor lunar parallax correction. The effect applied to characteristic declinations is shown graphically for the present time (J2000.0) in figure 12.11. For example, in a latitude of 30°, an object which goes through the zenith rises at azimuth 55°. Lunar major standstill risings occur at azimuths 56.7° and 123.5°, lunar minor standstill risings at azimuths 69° and 111°. The summer sun rises at 62.6°, the winter sun at 117.3°. An object which goes through the nadir rises at 125°.

The blue lines seem to vanish at $\varphi = 45°$: while there are still objects going through the zenith in higher latitudes, they are circumpolar and do not cross the horizon.

For the lunar events, there are two lines each drawn by the plugin, for maximum and minimum distance of the moon. The lunar extreme declinations are computed taking horizon parallax effects into account. For technical reasons however, the derived declinations are then used to draw *small circles* of constant declinations on the sphere, without taking the change of lunar horizontal parallax into account. Note that therefore the observed declination of the moon at the major standstill can exceed the indicated limits if it is high in the sky. The main purpose of this plugin is however to show an indication of the intersection of the standstill line with the horizon.

It may be very instructive to let the time run quite fast and observe the declination line of "current moon" swinging between its north and south limits each month. These limits grow and shrink between the Major and Minor Standstills in the course of 18.6 years.

The sun likewise swings between the solstices. Over centuries, the solstice declinations very slightly move as well due to the slightly changing obliquity of the ecliptic.

12.9.3 Azimuth Lines

Some religions, e.g., Islam or Bahai, adhere to a practice of observing a prayer direction towards a particular location. Azimuth lines for two locations can be shown, these lines indicate the great circle direction towards the locations which can be edited in the configuration window. Default locations are Mecca (Kaaba) and Jerusalem. The azimuth q towards a location $T = (\lambda_T, \varphi_T)$ are computed for an observer at $O = (\lambda_O, \varphi_O)$ based on spherical trigonometry on a spherical Earth [1]:

$$q = \arctan\left(\frac{\sin(\lambda_T - \lambda_O)}{\cos\varphi_O \tan\varphi_T - \sin\varphi_O \cos(\lambda_T - \lambda_O)}\right) \tag{12.2}$$

In addition, up to two vertical lines with arbitrary azimuth and custom label can be shown.

12.9.4 Configuration Options

The configuration dialog allows the selection of the lines which are of interest to you. In addition, you can select the color of the lines by clicking on the color swatches.[33]

Section (ArchaeoLines) in config.ini file

Apart from changing settings using the plugin configuration dialog, you can also edit config.ini file by yourself for changes of the settings for the ArchaeoLines plugin – just make it carefully!

ID	Type	Default
enable_at_startup	bool	false
color_equinox	float R,G,B	1.00,1.00,0.5
color_solstices	float R,G,B	1.00,1.00,0.25
color_crossquarters	float R,G,B	1.00,0.75,0.25
color_major_standstill	float R,G,B	0.25,1.00,0.25
color_minor_standstill	float R,G,B	0.20,0.75,0.20
color_zenith_passage	float R,G,B	1.00,0.75,0.75
color_nadir_passage	float R,G,B	1.00,0.75,0.75
color_selected_object	float R,G,B	1.00,1.00,1.00
color_current_sun	float R,G,B	1.00,1.00,0.75
color_current_moon	float R,G,B	0.50,1.00,0.50
color_current_planet	float R,G,B	0.25,0.80,1.00
color_geographic_location_1	float R,G,B	0.25,1.00,0.25
color_geographic_location_2	float R,G,B	0.25,0.25,1.00
color_custom_azimuth_1	float R,G,B	0.25,1.00,0.25

---

[33]Unfortunately, on Windows in OpenGL mode, the color dialog hides behind the Stellarium window when in fullscreen mode. So, before editing line colors, please leave fullscreen mode!

color_custom_azimuth_2	float R,G,B	0.25,0.50,0.75
show_equinox	bool	true
show_solstices	bool	true
show_crossquarters	bool	true
show_major_standstills	bool	true
show_minor_standstills	bool	true
show_zenith_passage	bool	true
show_nadir_passage	bool	true
show_selected_object	bool	true
show_current_sun	bool	true
show_current_moon	bool	true
show_current_planet	string	none
show_geographic_location_1	bool	false
show_geographic_location_2	bool	false
geographic_location_1_longitude	double	39.826175
geographic_location_1_latitude	double	21.4276
geographic_location_1_label	string	Mecca (Qibla)
geographic_location_2_longitude	double	35.235774
geographic_location_2_latitude	double	31.778087
geographic_location_2_label	string	Jerusalem
show_custom_azimuth_1	bool	false
show_custom_azimuth_2	bool	false
custom_azimuth_1_angle	double	0.0
custom_azimuth_2_angle	double	0.0
custom_azimuth_1_label	string	custAzi1
custom_azimuth_2_label	string	custAzi2

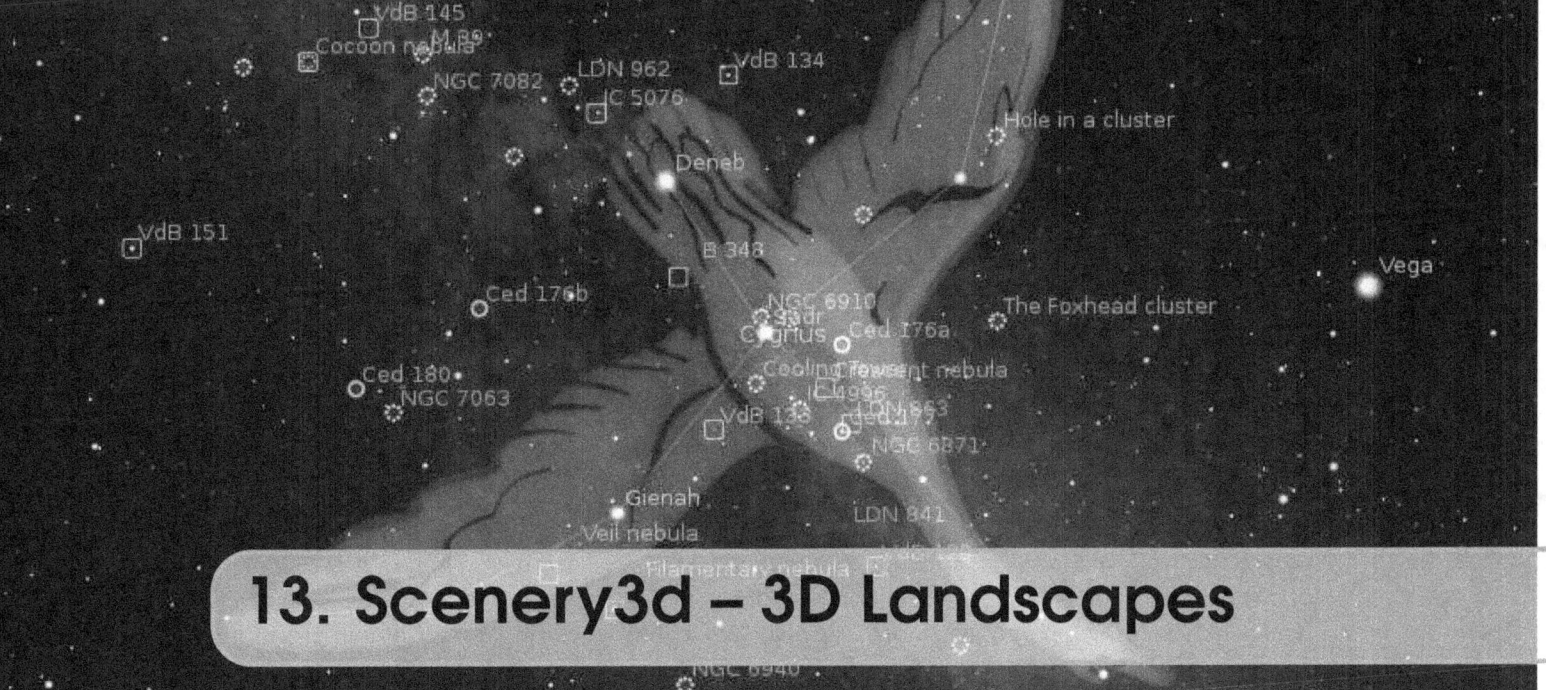

13. Scenery3d – 3D Landscapes

GEORG ZOTTI AND FLORIAN SCHAUKOWITSCH

13.1 Introduction

Have you ever wished to be able to walk through Stonehenge or other ancient building structures described as being constructed with astronomical orientation in mind, and experience such orientation in a 3D virtual environment that also provides a good sky simulation?

The Stellarium Scenery3d plugin allows you to see architectural 3D models embedded in a landscape combined with the excellent representation of the sky provided by Stellarium. You can walk around, check for (or demonstrate) possible astronomical alignments of ancient architecture, see sundials and other shadow casters in action, etc.

13.2 Usage

You activate the plugin with the *circular enclosure* button $\boxed{\bigcirc}$ at screen bottom or by pressing $\boxed{\text{Ctrl}}+\boxed{\text{W}}$. The other button with circular enclosure and tool icon $\boxed{\oslash}$ (or $\boxed{\text{Ctrl}}+\boxed{⇧}+\boxed{\text{W}}$) opens the settings dialog. Once loaded and displaying, you can walk around pressing $\boxed{\text{Ctrl}}$ plus cursor keys. Change eye height with $\boxed{\text{Ctrl}}+\boxed{\text{Page↑}}/\boxed{\text{Ctrl}}+\boxed{\text{Page↓}}$ keys. Adding $\boxed{⇧}$ key increases speed by 10, adding $\boxed{\text{Alt}}$ multiplies by 5 (pressing both keys multiplies by 50!). If you release $\boxed{\text{Ctrl}}$ before the cursor key, animation will continue. (Press $\boxed{\text{Ctrl}}$+any cursor key to stop moving.)

Further key bindings exist which can be configured using the Stellarium default key-binding interface. Some options are also available in the Scenery3d dialog. For example, coordinate display can be enabled with $\boxed{\text{Ctrl}}+\boxed{\text{R}}+\boxed{\text{T}}$. If your models are georeferenced in a true geographical coordinate grid, e.g. UTM or Gauss-Krueger, you will especially like this, and this makes the plugin usable for scientific purposes. Display shows grid name, Easting, Northing, Altitude of ground, and eye height above ground.

Other features include a virtual "torchlight", which can be enabled with $\boxed{\text{Ctrl}}+\boxed{\text{R}}+\boxed{\text{L}}$ to give additional local illumination around the viewer to help to see in the dark. Interesting points of view can be saved and restored later by the user, including a description of the view. Scene authors can also distribute predefined viewpoints in their scene.

The plugin also simulates the shadows of the scene's objects cast by the Sun, Moon and even Venus (only 1 shadow caster used at a time, you will never see shadows cast by Venus in moonlight), so you could use it for examining sundials, or analyze and simulate light-and-shadow interactions in archaeological structures.

13.3   Hardware Requirements & Performance

In order to work with the non-linear projection models in Stellarium, this plugin uses a trick to create the foreground renderings: it renders the scene into the six planes of a so-called cubemap, which is then correctly reprojected onto the sides of a cube, depending on the current projection settings. Your graphics card must be able to do this, i.e. it must support the OpenGL extension called EXT_framebuffer_object. Typical modern 3D cards (by NVidia or ATI/AMD) support this extension. In case your graphics hardware does not suppport it, the plugin will still work, but you are limited to perspective projection.

You can influence rendering quality, but also speed, using the plugin's GUI, which provides some options such as enabling the use of shadows, bumpmapping (provides more realistic surface lighting) or configuring the sizes of the textures used for the cubemap or shadowmaps. Larger values there improve the quality, but require faster hardware and more video memory for smooth results.

Because the "cubemap trick" requires quite a large amount of performance (in essence, the scene has to be rendered 6 times), there are some options available that try to reduce this burden. The first option is to change the type of the "cubemap". The most compatible setting is *6 textures*, which seems to work best on older integrated Intel GPUs. The recommended default is the second setting, *Cubemap*, which uses a more modern OpenGL feature and generally works a bit faster than *6 textures* on more modern graphics cards. Finally, the *Geometry shader* option tries to render all 6 cube faces at once. This requires a more recent GPU + drivers (at least OpenGL 3.2 must be supported), the setting is disabled otherwise. Depending on your hardware and the scene's complexity, this method may give a speedup or may be slower, you must find this out yourself.

Another option prevents re-rendering of the cubemap if nothing relevant has changed. You can define the interval (in Stellarium's simulation time) in which nothing is updated in the GUI. You can still rotate the camera without causing a re-draw, giving a subjective performance that is close to Stellarium's performance without Scenery3d. When moving, the cubemap will be updated. You can enable another option that only causes 1 or 2 sides of the cubemap to be updated while you move, giving a speedup but causing some parts of the image to be outdated and discontinuous. The cubemap will be completed again when you stop moving.

Shadow rendering may also cause quite a performance impact. The *Simple shadows* option can speed this up a lot, at the cost of shadow quality especially in larger scenes. Another performance/quality factor is shadow filtering. The sharpest (and fastest) possible shadows are achieved with filtering *Off*, but depending on shadowmap resolution and scene size the shadows may look quite "blocky". *Hardware* shadow filtering is usually very fast, but may not improve appearance a lot. Therefore, there are additional filter options available, the *High* filter option is relatively expensive. Finally, the *PCSS* option allows to approximate the increase of solar and lunar shadow penumbras relative to the distance from their shadow casters, i.e. shadows are sharp near contact points, and more blurred further away. This again requires quite a bit of performance, and only works if the shadow filter option is set to *Low* or *High* (without *Hardware*).

The configuration GUI shows tooltips for most of its settings, which can explain what a setting

Geometry	Yes
Lights	Yes
Clay	No
Photomatched	Yes
DefaultUVs	No
Instanced	No

Table 13.1: Kerkythea Export Settings

does. All settings are saved automatically, and restored when you reopen Stellarium.

13.3.1 Performance notes

On reasonably good hardware (tested on a notebook PC with NVidia M580 GTS), models with about 850.000 triangles are working nicely with shadows and bumpmaps. On very small hardware like single-board computers with native OpenGL ES2, models may be limited to 64k vertices (points). If display is too slow, switch to perspective projection: all other projections require almost sixfold effort! You should also prefer the "lazy" cubemap mode, where the scene is only rendered in specific timesteps or when movement happens.

13.4 Model Configuration

The model format supported in Scenery3d is Wavefront .OBJ, which is pretty common for 3D models. You can use several modeling programs to build your models. Software such as **Blender**, **Maya**, **3D Studio Max** etc. can export OBJ.

13.4.1 Exporting OBJ from Sketchup

A simple to use and cost-free modeling program is **Sketchup**, commonly used to create the 3D buildings seen in **Google Earth**. It can be used to create georeferenced models. OBJ is not a native export format for the standard version of **Sketchup**. If you are not willing to afford **Sketchup Pro**, you have to find another way to export a textured OBJ model.

One good exporter is available in the **Kerkythea** renderer project[1]. You need **SU2KT 3.17** or better, and **KT2OBJ 1.1.0** or better. Deselect any selection, then export your model to the **Kerkythea** XML format with settings shown in 13.1. (Or, with selection enabled, make sure settings are No-Yes-Yes-No-Yes-No-No.) You do not have to launch **Kerkythea** unless you want to create nice renderings of your model. Then, use the **KT2OBJ** converter to create an OBJ. You can delete the XML after the conversion. Note that some texture coordinates may not be exported correctly. The setting Photomatched:Yes seems now to have corrected this issue, esp. with distorted/manually shifted textures.

Another free OBJ exporter has been made available by TIG: OBJexporter.rb[2]. This is the only OBJ exporter tested so far capable of handling large TIN landscapes (> 450.000 triangles). As of version 2.6 it seems to be the best OBJ exporter available for **Sketchup**.

This exporter swaps Y/Z coordinates, but you can add a key to the config file to correct swapped axes, see below. Other exporters may also provide coordinates in any order of X, Y, Z – all those can be properly configured.

---

[1] Available at http://www.kerkythea.net/cms/

[2] Available from http://forums.sketchucation.com/viewtopic.php?f=323&t=33448

Another (almost) working alternative: `ObjExporter.rb` by author Honing. Here, export with settings 0xxx00. This will not create a `TX...` folder but dump all textures in the same directory as the OBJ and MTL files. Unfortunately, currently some material assignments seem to be bad.

Yet another exporter, `su2objmtl`, does also not provide good texture coordinates and cannot be recommended at this time.

13.4.2 Notes on OBJ file format limitations

The OBJ format supported is only a subset of the full OBJ format: Only (optionally textured) triangle meshes are supported, i.e., only lines containing statements: `mtllib`, `usemtl`, `v`, `vn`, `vt`, `f` (with three elements only!), `g`. Negative vertex numbers (i.e., a specification of relative positions) are not supported.

A further recommendation for correct illumination is that all vertices should have vertex normals. **Sketchup** models exported with the **Kerkythea** or TIG plugins should have correct normals. If your model does not provide them, default normals can be reconstructed from the triangle edges, resulting in a faceted look.

If possible, the model should also be triangulated, but the current loader may also work with non-triangle geometry. The correct use of objects (o) and groups (g) will improve performance: it is best if you pre-combine all objects that use the same material into a single one. The loader will try to optimize it anyways if this is not the case, but can do this only partly (to combine 2 objects with the same material into 1, it requires them to follow directly after each other in the OBJ). A simple guide to use **Blender**[3] for this task follows:

1. File ≫ Import ≫ Wavefront .obj - you may need to change the forward/up axes for correct orientation, try "-Y forward" and "Z up"
2. Select an object which has a shared material
3. Press ⇧ + L and select 'By Material'
4. Select 'Join' in the left (main) tool window
5. Repeat for other objects that have shared materials
6. Export the .obj, making sure to select the same forward/up axes as in the import, also make sure "Write Normals", "Write Materials" and "Include UVs" are checked

For transparent objects (with a `d` or `Tr` value, alpha testing does NOT need this), this recommendation does NOT hold: for optimal results, each separate transparent object should be exported as a separate "OBJ object". This is because they need to be sorted during rendering to achieve correct transparency. If the objects are combined already, you can separate them using **Blender**:

1. Import .obj (see above)
2. Select the combined transparent object
3. Enter "Edit" mode with ⇥ and make sure everything is selected (press A if not)
4. Press P and select "By loose parts", this should separate the object into its unconnected regions
5. Export .obj (see above), also check "Objects as OBJ Objects"

The MTL file specified by `mtllib` contains the material parameters. The minimum that should be specified is either `map_Kd` or a `Kd` line specifying color values used for the respective faces. But there are other options in MTL files, and the supported parameters and defaults are listed in Table 13.2.

If no ambient color is specified, the diffuse color values are taken for the ambient color. An optional emissive term `Ke` can be added, which is modulated to only be visible during nighttime. This also requires the landscape's self-illumination layer to be enabled. It allows to model self-illuminating objects such as street lights, windows etc. It can optionally also be modulated by the emissive texture `map_Ke`.

---

[3]`http://www.blender.org`

Parameter	Default	Range	Meaning
Ka	set to Kd values	0...1 each	R/G/B Ambient color
Kd	0.8 0.8 0.8	0...1 each	R/G/B Diffuse color
Ke	0.0 0.0 0.0	0...1 each	R/G/B Emissive color
Ks	0.0 0.0 0.0	0...1 each	R/G/B Specular color
Ns	8.0	$0...\infty$	shinyness
d or Tr	1.0	0...1	opacity
bAlphatest	0	0 or 1	perform alpha test
bBackface	0	0 or 1	render backface
map_Kd	(none)	filename	texture map to be mixed with Ka, Kd
map_Ke	(none)	filename	texture map to be mixed with Ke
map_bump	(none)	filename	normal map for surface roughness
illum	2	integer	illumination mode in the standard MTL format.

Table 13.2: MTL parameters evaluated

If a value for Ks is specified, specularity is evaluated using the Phong reflection model[4] with Ns as the exponential shininess constant. Larger shininess means smaller specular highlights (more metal-like appearance). Specularity is not modulated by the texture maps. Unfortunately, some 3D editors export unusable default value combinations for Ks and Ns. **Blender** may create lines with Ks=1/1/1 and Ns=0. This creates a look of "partial overexposed snow fields". While the values are allowed in the specification, in most cases the result looks ugly. *Make sure to set Ns to 1 or higher, or disable those two lines.*

If a value for d or Tr exists, alpha blending is enabled for this material. This simulates transparency effects. Transparency can be further controlled using the alpha channel of the map_Kd texture.

A simpler and usually more performant way to achieve simple "cutout" transparency effects is alpha-testing, by setting bAlphatest to 1. This simply discards all pixels of the model where the alpha value of the map_Kd is below the transparency_threshold value from scenery3d.ini, making "holes" in the model. This also produces better shadows for such objects. If required, alpha testing can be combined with "real" blending-based transparency.

Sometimes, exported objects only have a single side ("paper wall"), and are only visible from one side when looked at in Scenery3d. This is caused by an optimization called back-face culling, which skips drawing the back sides of objects because they are usually not visible anyway. If possible, avoid such "thin" geometry, this will also produce better shadows on the object. As a workaround, you can also set bBackface to 1 to disable back-face culling for this material.

The optional map_bump enables the use of a tangent-space normal maps[5], which provides a dramatic improvement in surface detail under illumination.

13.4.3 Configuring OBJ for Scenery3d

The walkaround in your scene can use a ground level (piece of terrain) on which the observer can walk. The observer eye will always stay "eye height" above ground. Currently, there is no collision detection with walls implemented, so you can easily walk through walls, or jump on high towers, if their platform or roof is exported in the ground layer. If your model has no explicit ground layer,

---

[4]https://en.wikipedia.org/wiki/Phong_reflection_model
[5]https://en.wikipedia.org/wiki/Normal_mapping

walk will be on the highest surface of the scenery layer. If you use the special name NULL as ground layer, walk will be above zero_ground_height level.

Technically, if your model has cavities or doors, you should export your model twice. Once, just the ground plane, i.e. where you will walk. Of course, for a temple or other building, this includes its socket above soil, and any steps, but pillars should not be included. This plane is required to compute eye position above ground. Note that it is not possible to walk in several floors of a building, or in a multi-plane staircase. You may have to export several "ground" planes and configure several scenery directories for those rare cases. For optimal performance, the ground model should consist of as few triangles as you can tolerate.

The second export includes all visible model parts, and will be used for rendering. Of course, this requires the ground plane again, but also all building elements, walls, roofs, etc.

If you have not done so by yourself, it is recommended to separate ground and buildings into Sketchup layers (or similar concepts in whichever editor you are using) in order to easily switch the model to the right state prior to exporting.

Filename recommendations:

`<Temple>.skp`	Name of a Sketchup Model file. (The <> brackets signal "use your own name here!") The SKP file is not used by Scenery3d, but you may want to leave it in the folder for later improvements.
`<Temple>.obj`	Model in OBJ format.
`<Temple>_ground.obj`	Ground layer, if different from Model file.

OBJ export may also create folders TX_<Temple> and TX_<Temple>_ground. You can delete the TX_<Temple>_ground folder, <Temple>_ground.obj is just used to compute vertical height.

Put the OBJ, MTL and TX directories into a subdirectory of your user directory (see section 5.1), e.g. <USERDATA>/Stellarium/scenery3d/<Temple>, and add a text file into it called scenery3d.ini (This name is mandatory!) with content described as follows.

```
[model]
name=<Temple>
```
Unique ID within all models in scenery3d directory. Recommendation: use directory name.

`landscape=<landscapename>`   Name of an available Stellarium landscape.

This is required if the landscape file includes geographical coordinates and your model does not: First, the location coordinates of the landscape.ini file are used, then location coordinates given here. The landscape also provides the background image of your scenery. If you want a zero-height (mathematical) horizon, use the provided landscape called Zero Horizon.

`scenery=<Temple>.obj`	The complete model, including visible ground.
`ground=<Temple>_ground.obj`	Optional: separate ground plane. (NULL for zero altitude.)
`description=<Description>`	A basic scene description (including HTML tags)

The scenery3d.ini may contain a simple scene description, but it is recommended to use the *localizable* description format: in the scene's directory (which contains scenery3d.ini) create files in the format description.<lang>.utf8 which can contain arbitrary UTF-8–encoded HTML content. <lang> stands for the ISO 639 language code.

```
author=<Your Name yourname@yourplace.com>
copyright=<Copyright Info>
```

`obj_order=XYZ`	Use this if you have used an exporter which swaps Y/Z coordinates. Defaults to XYZ, other options: XZY, YZX, YXZ, ZXY, ZYX
`camNearZ=0.3`	This defines the distance of the camera near plane, default 0.3. Everything closer than this value to the camera can not be displayed. Must be larger than zero. It may seem tempting to set this very small, but this will lead to accuracy issues. Recommendation is not to go under 0.1
`camFarZ=10000`	Defines the maximal viewing distance, default 10000.
`shadowDistance=<val>`	The maximal distance shadows are displayed. If left out, the value from `camFarZ` is used here. If this is set to a smaller value, this may increase the quality of the shadows that are still visible.
`shadowSplitWeight=0..1`	Decimal value for further shadow tweaking. If you require better shadows up close, try setting this to higher values. The default is calculated using a heuristic that incorporates scene size.

`[general]`

The general section defines some further import/rendering options.

`transparency_threshold=0.5`	Defines the alpha threshold for alpha-testing, as described in section 13.4.2. Default `0.5`
`scenery_generate_normals=0`	Boolean, if true normals are recalculated by the plugin, instead of imported. Default `false`
`ground_generate_normals=0`	Boolean, same as above, for ground model. Default `false`.

`[location]`

Optional section to specify geographic longitude λ, latitude φ, and altitude. The secton is required if `coord/convergence_angle=from_grid`, else location is inherited from landscape.

```
planet = Earth
latitude = +48d31'30.4"
longitude = +16d12'25.5"
altitude =from_model|<int>
```

`latitude = +48d31'30.4"`	Required if `coord/convergence_angle=from_grid`
`longitude = +16d12'25.5"`	`"-"`
`altitude =from_model|<int>`	altitude (for astronomical computations) can be computed from the model: if `from_model`, it is computed as $(z_{min} + z_{max})/2 + $ `orig_H`, i.e. from the model bounding box centre height.

```
display_fog = 0
atmospheric_extinction_coefficient = 0.2
atmospheric_temperature = 10.0
atmospheric_pressure = -1
light_pollution = 1
[coord]
```

Entries in the `[coord]` section are again optional, default to zero when not specified, but are required if you want to display meaningful eye coordinates in your survey (world) coordinate system, like UTM or Gauss-Krüger.

`grid_name=<string>`	Name of grid coordinates, e.g. `"UTM 33 U (WGS 84)"`, `"Gauss-Krüger M34"` or `"Relative to <Center>"` This name is only displayed, there is no evaluation of its contents.
`orig_E=<double>` | (Easting)	East-West-distance to zone central meridian

```
orig_N=<double>  |  (Northing)    North distance from Equator
orig_H=<double>  |  (Height)      Altitude above Mean Sea Level of model origin
```

These entries describe the offset, in metres, of the model coordinates relative to coordinates in a geographic grid, like Gauss-Krüger or UTM. If you have your model vertices specified in grid coordinates, do not specify `orig_...` data, but please definitely add `start_...` data, below.

Note that using grid coordinates without offset for the vertices is usually a bad idea for real-world applications like surveyed sites in UTM coordinates. Coordinate values are often very large numbers (ranging into millions of meters from equator and many thousands from the zone meridian). If you want to assign millimetre values to model vertices, you will hit numerical problems with the usual single-precision floating point arithmetic. Therefore we can specify this offset which is only necessary for coordinate display.

Typically, digital elevation models and building structures built on those are survey-grid aligned, so true geographical north for a place with geographical longitude λ and latitude φ will in general not coincide with grid north, the difference is known as *meridian convergence*[6].

$$\gamma(\lambda, \varphi) = \arctan(\tan(\lambda - \lambda_0) \sin \varphi) \tag{13.1}$$

This amount can be given in `convergence_angle` (degrees), so that your model will be rotated clockwise by this amount around the vertical axis to be aligned with True North[7].

```
convergence_angle=from_grid|<double>
grid_meridian=<double>|+<int>d<int>'<float>"
```

`grid_meridian` is the central meridian λ_0 of grid zone, e.g. for UTM or Gauss-Krüger, and is only required to compute convergence angle if `convergence_angle=from_grid`.

```
zero_ground_height=<double>
```

height of terrain outside `<Temple>_ground.OBJ`, or if ground=NULL. Allows smooth approach from outside. This value is relative to the model origin, or typically close to zero, i.e., use a Z value in model coordinates, not world coordinates! (If you want the terrain height surrounding your model to be `orig_H`, use 0, not the correct mean height above sea level!) Defaults to minimum of height of ground level (or model, resp.) bounding box.

```
start_E=<double>
start_N=<double>
start_H=<double>        only meaningful if ground==NULL, else H is derived from ground
start_Eye=<double>   default: 1.65m
start_az_alt_fov=<az_deg>,<alt_deg>,<fov_deg>
                     initial view direction and field of view.
```

`start_...` defines the view position to be set after loading the scenery. Defaults to center of model boundingbox.

It is advisable to use the grid coordinates of the location of the panoramic photo (landscape) as `start_...` coordinates, or the correct coordinates and some carefully selected `start_az_alt_fov` in case of certain view corridors (temple axes, ...).

---

[6]http://en.wikipedia.org/wiki/Transverse_Mercator_projection

[7]Note that Sketchup's *georeferencing dictionary* provides a NorthAngle entry, which is $360 -$ convergence_angle.

13.4.4 Concatenating OBJ files

Some automated workflows may involve tiled landscape areas, e.g. to overcome texture limitations or triangle count limits in simpler tools like **Sketchup**. In this case you can create separate meshes in the same coordinate system, but you need to concatenate them. One powerful program to assemble your parts is again **Blender**.

In **Blender**, import the OBJ files File ⟫ Import ⟫ Wavefront .obj . If your OBJ coordinates have Z as vertical axis (common for terrestrial models), use "Z up", "-Y Forward" as import settings for the coordinate axes. The model will appear south-up. Switch on textured view if required (Alt + Z). If the scene looks almost black now, reconfigure the light to be sunlight. If you have created a VRML of some structures in **ArcScene**, export that without shifting to center, and import the WRL file in **Blender** with default orientation "Y up", "Z forward". Models from other sources may still be different. In the end, all parts should fit neatly together. **Blender** is a very powerful program, you can enhance your model as you wish.

In the end, select all relevant parts which you want to have visible in Stellarium (click on the lines in the outline view containing the object names to light them up gray, then Rightclick-Select) and press Ctrl + J to optionally join them, then export to a single OBJ File ⟫ Export ⟫ Wavefront .obj . In the export options, apply "Selection Only", and "-Y Forward" and "Z Up"[8].
Verify the new model loads correctly, e.g. in **Meshlab**[9]!

13.4.5 Working with non-georeferenced OBJ files

There exists modeling software which produces nice models, but without concept of georeference. One spectacular example is **AutoDesk PhotoFly**, a cloud application which delivers 3D models from a bunch of photos uploaded via its program interface. This "technological preview" is in version 2 and free of cost as of mid-2011.

The problem with these models is that you cannot assign surveyed coordinates to points in the model, so either you can georeference the models in other applications, or you must find the correct transformation matrix. Importing the OBJ in **Sketchup** may take a long time for detailed photo-generated models, and the texturing may suffer, so you can cut the model down to the minimum necessary e.g. in **Meshlab**, and import just a stub required to georeference the model in **Sketchup**.

Now, how would you find the proper orientation? The easiest chance would be with a structure visible in the photo layer of **Google Earth**. So, start a new model and immediately add location from the **Google Earth** interface. Then you can import the OBJ with TIG's importer plugin. If the imported model looks perfect, you may just place the model into the **Sketchup** landscape and export a complete landscape just like above. If not, or if you had to cut/simplify the OBJ to be able to import it, you can rotate/scale the OBJ (it must be grouped!). If you see a shadow in the photos, you may want to set the date/time of the original photos in the scene and verify that the shadows created by **Sketchup** illuminating the model match those in the model's photo texture. When you are satisfied with placement/orientation, you create a scenery3d.ini like above with the command Plugins ⟫ ASTROSIM/Stellarium scenery3d helpers ⟫ Create scenery3d.ini .

Then, you select the OBJ group, open the Windows ⟫ Ruby Console and readout data by calling Plugins ⟫ ASTROSIM/Stellarium scenery3d helpers ⟫ Export transformation of selected group .

On the Ruby console, you will find a line of numbers (the 4×4 transformation matrix) which you copy/paste (all in one line!) into the [model] section in scenery3d.ini.

```
obj2grid_trafo=<a11>,<a12>,<a13>,<a14>,<a21>,<a22>,<a23>,<a24>,
               <a31>,<a32>,<a33>,<a34>,<a41>,<a42>,<a43>,<a44>
```

---

[8]http://blender.stackexchange.com/questions/3352/merging-multiple-obj-files
[9]http://www.meshlab.org

You edit the `scenery3d.ini` to use your full (unmodified) **PhotoFly** model and, if you don't have a panorama, take `Zero Horizon` landscape as (no-)background. It depends on the model if you want to be able to step on it, or to declare ground=NULL for a constant-height ground. Run Stellarum once and adjust the `start_N`, `start_E` and `zero_ground_height`.

13.4.6 Rotating OBJs with recognized survey points

If you have survey points measured in a survey grid plus a photomodel with those points visible, you can use Meshlab to find the model vertex coordinates in the photo model, and some other program like **CoordTrans** in the **JavaGraticule3D** suite to find either the matrix values to enter in `scenery3d.ini` or even rotate the OBJ points. However, this involves more math than can be described here; if you came that far, you likely know the required steps. Here it really helps if you know how to operate automatic text processors like **AWK**.

13.5 Predefined views

You can also configure and distribute some predefined views with your model in a `viewpoints.ini` file. The viewpoints can be loaded and stored with the viewpoint dialog which you can call with the

[⊚] button. See the provided "Sterngarten" scene for an example. These entries are not editable by the user through the interface. The user can always save his own views, they will be saved into the file `userviews.ini` in the user's Stellarium user directory, and are editable.

```
[StoredViews]
size=<int>                 Defines how many entries are in this file.
                           Prefix each entry with its index!
1/label=<string>           The name of this entry
1/description=<string>     A description of this entry (can include HTML)
1/position=<x,y,z,h>       The x,y,z grid coordinates
                           (like orig_* or start_* in scenery3d.ini)
                           + the current eye height h
1/view_fov=<az_deg,alt_deg,fov_deg>  The view direction + FOV
                                     (like start_az_alt_fov in scenery3d.ini)
; an example for a second entry (note the 2 at the beginning of each line!)
2/label       = Signs
2/description = Two signs that describe the Sterngarten
2/position    = 593155.2421,5333348.6304,325.7295809038,0.8805
2/view_fov    = 84.315399,-8.187078,83.000000
```

Authors and Acknowledgements

Scenery3d was conceived by Georg Zotti for the ASTROSIM[10] project. A first prototype was implemented in 2010/2011 by Simon Parzer and Peter Neubauer as student work supervised by Michael Wimmer (TU Wien). Models for accuracy tests (Sterngarten, Testscene), and later improvements in integration, user interaction, `.ini` option handling, OBJ/MTL loader bugfixes and georeference testing by Georg Zotti.

Andrei Borza in 2011/12 further improved rendering quality (shadow mapping, normal mapping) and speed [70, 73, 74].

In 2014/15, Florian Schaukowitsch adapted the code to work with Qt 5 and the Stellarium 0.13 codebase, replaced the renderer with a more efficient, fully shader-based system, implemented various performance, quality and usability enhancements, and did some code cleanup. Both Andrei and Florian were again supervised by Michael Wimmer [71, 72].

---

[10]http://astrosim.univie.ac.at

This work has been originally created during the ASTROSIM project supported 2008-2012 by the Austrian Science Fund (FWF) under grant number P 21208-G19.

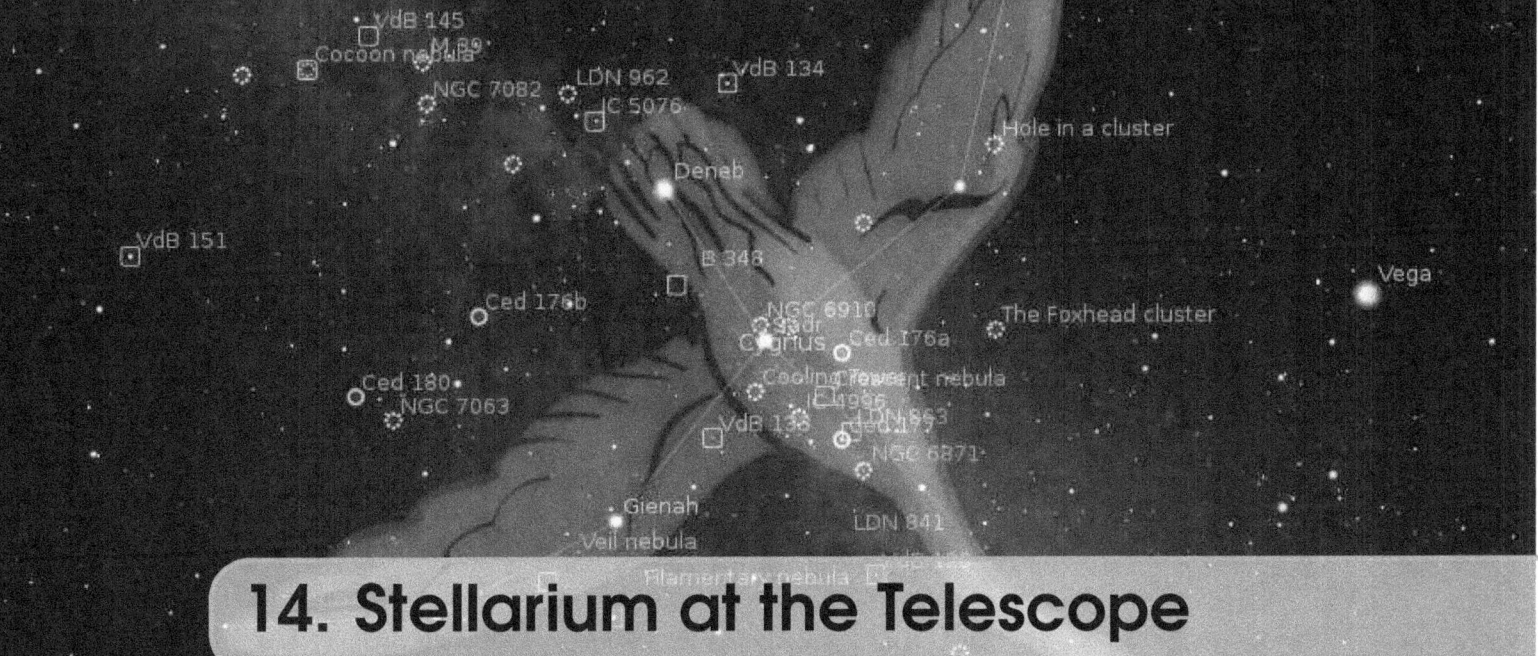

14. Stellarium at the Telescope

Stellarium is great for indoor use on the desktop, but it is also very useful outdoors under the real sky, and several plugins enhance its usability particularly for observers.

Two plugins are bundled with Stellarium which are designed to be used at the telescope: Oculars (section 14.1), which provides field of view hints for telescopes, oculars and sensors, and TelescopeControl (section 14.2), which allows you to send GOTO commands to many motorized telescopes. Other goto telescopes are supported by an external plugin which you must install separately: StellariumScope (section 14.3).

In addition, the Observability plugin (section 14.5) can be used for planning the best times to observe your favourite objects.

14.1 Oculars Plugin

This places a window on the screen that corresponds to the view through a telescope or on a camera. It reads from an editable data base.

When this plug in is active a circular view will appear around the selected object depicting what would be seen by the viewing object. On the top right hand side of the screen a menu will appear that can be used to select the viewing device, e.g., Camera, Eyepiece, Barlow type lenses, etc. This menu is filled with items from the `ocular.ini` file in the `modules/oculars` folder. This file can be edited from the Plugin's menu screen or with a text editor.

14.2   TelescopeControl Plugin

This plugin provides a simple control mechanism for motorised telescope mounts. The user selects an object (i.e. by clicking on something – a planet, a star etc.) and presses the telescope go-to key, and the telescope will be guided to the object.

Multiple telescopes may be controlled simultaneously.

WARNING

Stellarium cannot prevent your telescope from being pointed at the Sun. It is up to you to ensure proper filtering and safety measures are applied!

Never point your telescope at the Sun without a proper solar filter installed. The powerful light amplified by the telescope WILL cause *irreversible damage* to your eyes and/or your equipment.

Even if you don't do it deliberately, a slew during daylight hours may cause your telescope to point at the sun on its way to the given destination, so it is strongly recommended to avoid using the telescope control feature before sunset without appropriate protection.

14.2.1   Abilities and limitations

This plug-in allows Stellarium to send only 'slew' ('go to') commands to the device and to receive its current position. It cannot issue any other commands, so users should be aware of the possibility for mount collisions and similar situations. (To abort a slew, you can start another one to a safe position.)

Currently this plug-in does not allow satellite tracking, and is not very suitable for lunar or planetary observations.

14.2.2   Using this plug-in

Here are two general ways to control a device with this plug-in, depending on the situation:

DIRECT CONNECTION A device supported by the plug-in is connected with a cable to the computer running Stellarium

INDIRECT CONNECTION

    local A device is connected to the same computer but it is driven by a stand-alone telescope server program or a third-party application that can 'talk' to Stellarium;

    remote A device is connected to a remote computer and the software that drives it can 'talk' to Stellarium over the network; this software can be either one of Stellarium's stand-alone telescope servers, or a third party application.

Most older telescopes use cables that connect to a serial port (RS-232), the newer ones use USB (Universal Serial Bus). On Linux and Max OS X, both cases are handled identically by the plug-in. On Windows, a USB connection may require a 'virtual serial port' software, if it is not supplied with the cable or the telescope. Such a software creates a virtual ('fake') COM port that corresponds to the real USB port so it can be used by the plug-in. On all three platforms, if the computer has no 'classic' serial ports and the telescope can connect only to a serial port, a serial-to-USB (RS-232-to-USB) adapter may be necessary.

Telescope set-up (setting geographical coordinates, performing alignment, etc.) should be done before connecting the telescope to Stellarium.

14.2.3   Main window ('Telescopes')

The plug-in's main window can be opened:

- By pressing the | configure | button for the plug-in in the | Plugins | tab of Stellarium's Configuration window (opened by pressing | F2 | or the | button in the left toolbar).

- By pressing the $\boxed{\textsf{Configure telescopes...}}$ button in the $\boxed{\textsf{Slew to}}$ window (opened by pressing $\boxed{\textsf{Ctrl}}$ + $\boxed{\textsf{0}}$ or the respective button on the bottom toolbar).

The $\boxed{\textsf{Telescopes}}$ tab displays a list of the telescope connections that have been set up:

- The *number (#)* column shows the number used to control this telescope. For example, for telescope #2, the shortcut is $\boxed{\textsf{Ctrl}}$ + $\boxed{\textsf{2}}$.
- The *Status* column indicates if this connection is currently active or not. Unfortunately, there are some cases in which 'Connected' is displayed when no working connection exists.
- The *Type* field indicates what kind of connection is this:

  virtual means a virtual telescope

  local, Stellarium means a DIRECT connection to the telescope (see above)

  local, external means an INDIRECT connection to a program running on the same computer

  remote, unknown means an INDIRECT connection over a network to a remote machine.

To set up a new telescope connection, press the $\boxed{\textsf{Add}}$ button. To modify the configuration of an existing connection, select it in the list and press the $\boxed{\textsf{Configure}}$ button. In both cases, a telescope connection configuration window will open.

14.2.4   Telescope configuration window

Connection type

The topmost field represents the choice between the types of connections (see section 14.2.2): Telescope controlled by:

Stellarium, directly through a serial port is the DIRECT case

External software or a remote computer is the INDIRECT case

Nothing, just simulate one (a moving reticle) is a *virtual telescope* (no connection)

Telescope properties

Name is the label that will be displayed on the screen next to the telescope reticle.

Connection delay If the movement of the telescope reticle on the screen is uneven, you can try increasing or decreasing this value.

Coordinate system Some Celestron telescopes have had their firmware updated and now interpret the coordinates they receive as coordinates that use the equinox of the date (EOD, also known as JNow), making necessary this setting.

Start/connect at startup Check this option if you want Stellarium to attempt to connect to the telescope immediately after it starts. Otherwise, to start the telescope, you need to open the main window, select that telescope and press the $\boxed{\textsf{Start/Connect}}$ button.

Device settings

This section is active only for DIRECT connections (see above).

Serial port sets the serial port used by the telescope. There is a pop-up box that suggests some default values:

- On Windows, serial ports COM1 to COM10
- On Linux, serial ports /dev/ttyS0 to /dev/ttyS3 and USB ports /dev/ttyUSB0 to /dev/ttyUSB3
- On Mac OS X, the list is empty as it names its ports in a peculiar way. If you are using an USB cable, the default serial port of your telescope most probably is not in the list of suggestions. To list all valid serial port names in Mac OS X, open a terminal and type:

```
ls /dev/*
```

This will list all devices, the full name of your serial port should be somewhere in the list (for example, /dev/cu.usbserial-FTDFZVMK).

Device model : see 14.2.5 Supported devices.

Connection settings

Both fields here refer to INDIRECT connections, which implies communication over a network (TCP/IP).

Host can be either a host name or an IPv4 address such as '127.0.0.1'. The default value of 'localhost' means 'this computer'.

Modifying the default host name value makes sense only if you are attempting a remote connection over a network. In this case, it should be the name or IP address of the computer that runs a program that runs the telescope.

Port refers to the TCP port used for communication. The default value depends on the telescope number and ranges between 10001 and 10009.

Both values are ignored for DIRECT connections.

User Interface Settings: Field of view indicators

A series of circles representing different fields of view can be added around the telescope marker. This is a relic from the times before the Oculars plug-in (see 14.1) existed.

Activate the 'Use field of view indicators' option, then enter a list of values separated with commas in the field below. The values are interpreted as degrees of arc.

These marks can be used in combination with a virtual telescope to display a moving reticle with the Telrad circles.

'Slew telescope to' window

The $\boxed{\text{Slew telescope to}}$ window can be opened by pressing $\boxed{\text{Ctrl}}$ + $\boxed{\text{0}}$ or the respective button in the bottom toolbar.

It contains two fields for entering celestial coordinates, selectors for the preferred format (Hours-Minutes-Seconds, Degrees-Minutes-Seconds, or Decimal degrees), a drop-down list and two buttons.

The drop-down list contains the names of the currently connected devices. If no devices are connected, it will remain empty, and the $\boxed{\text{Slew}}$ button will be disabled.

Pressing the $\boxed{\text{Slew}}$ button slews the selected device to the selected set of coordinates. See the section about keyboard commands below for other ways of controlling the device.

Pressing the $\boxed{\text{Configure telescopes...}}$ button opens the main window of the plug-in.

TIP: Inside the 'Slew' window, underlined letters indicate that pressing $\boxed{\text{Alt}}$ + $\boxed{\text{underlined letter}}$ can be used instead of clicking. For example, pressing $\boxed{\text{Alt}}$ + $\boxed{\text{S}}$ is equivalent to clicking the $\boxed{\text{Slew}}$ button, pressing $\boxed{\text{Alt}}$ + $\boxed{\text{E}}$ switches to decimal degree format, etc.

Sending commands

Once a telescope is successfully started/connected, Stellarium displays a telescope reticle labelled with the telescope's name on its current position in the sky. The reticle is an object like every other in Stellarium - it can be selected with the mouse, it can be tracked and it appears as an object in the 'Search' window.

To point a device to an object: Select an object (e.g. a star) and press the number of the device while holding down the $\boxed{\text{Ctrl}}$ key. (For example, $\boxed{\text{Ctrl}}$ + $\boxed{\text{1}}$ for telescope #1.) This will move the telescope to the selected object.

To point a device to the center of the view: Press the number of the device while holding down the Alt key. (For example, $\boxed{\text{Alt}}$ + $\boxed{\text{1}}$ for telescope #1.) This will slew the device to the point in the center of the current view. (If you move the view after issuing the command, the target won't change unless you issue another command.)

To point a device to a given set of coordinates: Use the Slew to window (press Ctrl + 0).

14.2.5 Supported devices

All devices listed in the 'Device model' list are convenience definitions using one of the two built-in interfaces: the Meade LX200 (the Meade Autostar controller) interface and the Celestron NexStar interface.

The device list contains the following:

Celestron NexStar (compatible) Any device using the NexStar interface.

Losmandy G-11 A computerized telescope mount made by Losmandy (Meade LX-200/Autostar interface).

Meade Autostar compatible Any device using the LX-200/Autostar interface.

Meade ETX-70 (#494 Autostar, #506 CCS) The Meade ETX-70 telescope with the #494 Autostar controller and the #506 Connector Cable Set. According to the tester, it is a bit slow, so its default setting of 'Connection delay' is 1.5 seconds instead of 0.5 seconds.

Meade LX200 (compatible) Any device using the LX-200/Autostar interface.

Sky-Watcher SynScan AZ mount The Sky-Watcher SynScan AZ GoTo mount is used in a number of telescopes.

Sky-Watcher SynScan (version 3 or later) SynScan is also the name of the hand controller used in other Sky-Watcher GoTo mounts, and it seems that any mount that uses a SynScan controller version 3.0 or greater is supported by the plug-in, as it uses the NexStar protocol.

Wildcard Innovations Argo Navis (Meade mode) Argo Navis is a 'Digital Telescope Computer' by Wildcard Innovations. It is an advanced digital setting circle that turns an ordinary telescope (for example, a dobsonian) into a 'Push To" telescope (a telescope that uses a computer to find targets and human power to move the telescope itself). Just don't forget to set it to Meade compatibility mode and set the baud rate to 9600B1.

Virtual telescope

If you want to test this plug-in without an actual device connected to the computer, choose "Nothing, just simulate one (a moving reticle)" in the Telescope controlled by: field. It will show a telescope reticle that will react in the same way as the reticle of a real telescope controlled by the plug-in. See the section above about field of view indicators for a possible practical application (emulating 'Telrad' circles).

14.3　StellariumScope plugin

StellariumScope is a free add-on that enables you to control your telescope with Stellarium.

Features

- Provides an interface between Stellarium and the ASCOM telescope drivers.
- Provides the ability to both "Sync" and "Slew" the telescope. It's also possible to issue a stop/cancel command from Stellarium.
- You can easily host Stellarium on one computer linked to another control computer that hosts the telescope driver.
- The installation program will automatically install the documentation, but the link to the documentation is provided by developer[1] so you can read it before installation.
- There are earlier releases still available on the downloads page on Welsh Dragon Computing site.

The original StellariumScope program was designed and implemented by Scott of ByteArts and is still available for download[2]. If you have difficulties with the releases available on the Welsh Dragon Computing site[3], you may want to consider using the original version.

Figure 14.3 shows the interface and some of the options. Use this application (like all software that controls your mount) with supervision of your mount's movements.

14.4　Other telescope servers and Stellarium

Other developers have also been busy creating hard- and software often involving Arduino or Raspberry Pi boards which can control GOTO or PUSHTO (manually driven but position-aware, usually Dobsonian) telescopes and are ultimately controlled from Stellarium. Those are not related nor authored by the Stellarium team, so while we welcome such development (esp. open-sourced) in general, we cannot provide documentation nor any support.

A few examples:

iTelescope `http://simonbox.info/index.php/astronomy/93-raspberry-pi-itelescope`
node-telescope-server `https://www.npmjs.com/package/node-telescope-server`

One anonymous user sent a troubleshooting solution when connecting Stellarium to the Celestron **NexRemote** software:

> This involves connecting Stellarium to the **NexRemote** software controlling a Celestron NexStar telescope.
>
> One tricky Window XP issue I fixed was that my older laptop would transiently lose connection with Stellarium although the status would still be "Connected" and all looked normal.
>
> 3 (or whatever) slews would work. Next – nothing. Although all seemed well.
>
> I boosted the *NexRemote.exe* process in Windows XP to *High* under *Set Priority* under the **Windows Task Manager** via `Ctrl`+`Alt`+`Del`.
>
> All slews now proceed normally. Problem went away.[4]

---

[1]StellariumScope User's Guide — `http://welshdragoncomputing.ca/x/st/misc/stellariumscope_user_guide.2015.10.24.pdf`

[2]`http://www.bytearts.com/stellarium/`

[3]`http://welshdragoncomputing.ca/x/index.php/home/stellariumscope/about-stellariumscope`

[4]`https://sourceforge.net/p/stellarium/discussion/278769/thread/16e4c054/?limit=25#8ffa`

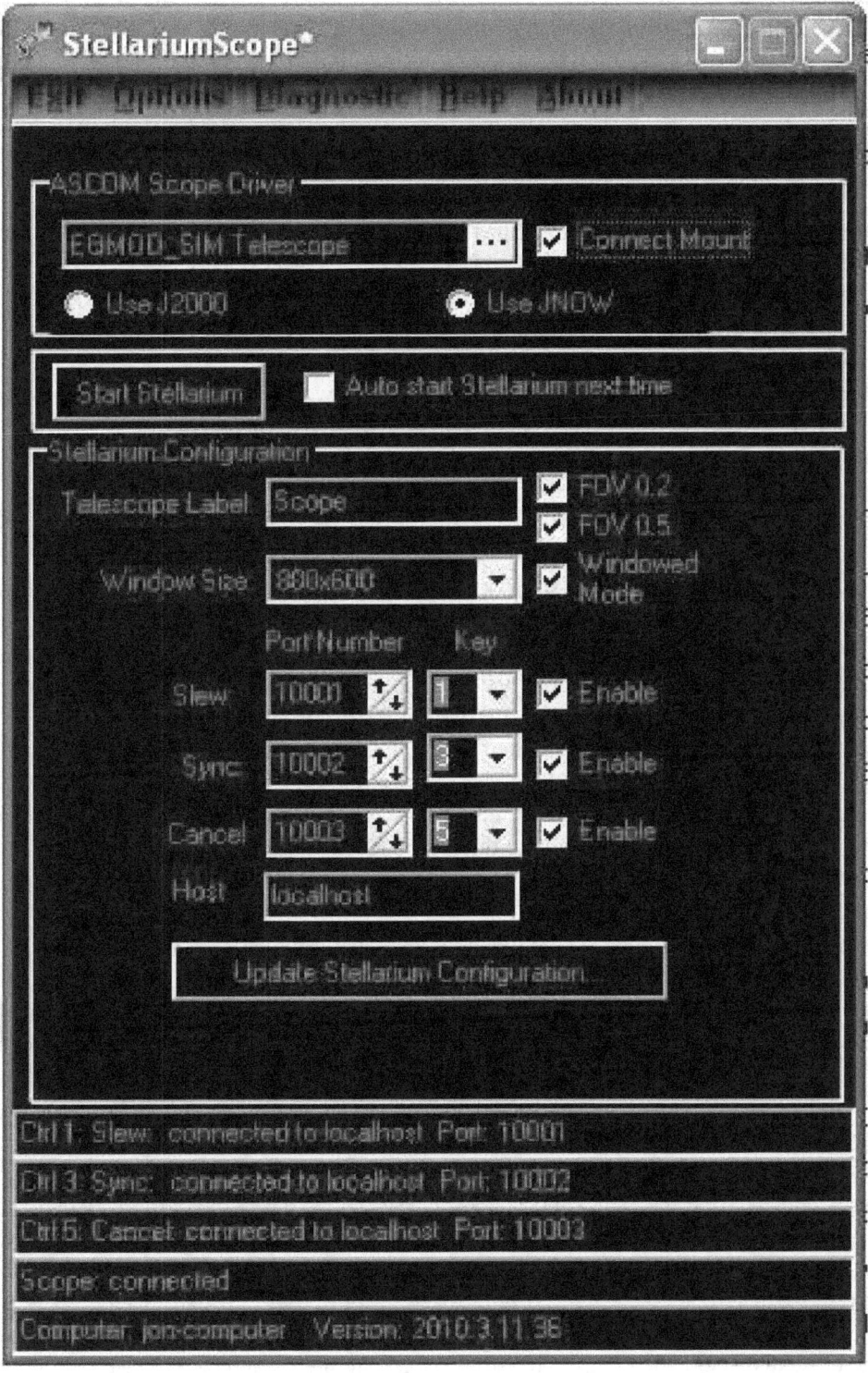

Figure 14.1: StellariumScope interface

14.5 Observability Plugin

This Plugin analyzes the observability of the selected object (or the screen center, if no object is selected). The plugin can show rise, transit, and set times, as well as the best epoch of the year (i.e., largest angular separation from the Sun), the date range when the source is above the horizon at dark night, and the dates of Acronychal and Cosmical rise/set. Ephemerides of the Solar-System objects and parallax effects are taken into account.

Explanation of some parameters

Sun altitude at twilight Any celestial object will be considered visible when the Sun is below this altitude. The altitude at astronomical twilight ranges usually between -12 and -18 degrees. This parameter is only used for the estimate of the range of observable epochs (see below).

Horizon altitude Minimum observable altitude (due to mountains, buildings, or just a limited telescope mount).

Today ephemeris Self-explanatory. The program will show the rise, set, and culmination (transit) times. The exact times for these ephemeris are given in two ways: as time spans (referred to the current time) and as clock hours (in local time).

Acronychal/Cosmical/Heliacal rise/set The days of Cosmical rise/set of an object are estimated as the days when the object rises (or sets) together with the rise/set of the Sun. The exact dates of these ephemeris depend on the Observer's location. On the contrary, the Acronycal rise (or set) happens when the star rises/sets with the setting/rising of the Sun (i.e., opposite to the Sun). On the one hand, it is obvious that the source is hardly observable (or not observable at all) in the dates between Cosmical set and Cosmical rise. On the other hand, the dates around the Acronychal set and rise are those when the altitude of the celestial object uses to be high when the Sun is well below the horizon (hence the object can be well observed). The date of Heliacal rise is the first day of the year when a star becomes visible. It happens when the star is close to the eastern horizon roughly before the end of the astronomical night (i.e., at the astronomical twilight). In the following nights, the star will be visibile during longer periods of time, until it reaches its Heliacal set (i.e., the last night of the year when the star is still visible). At the Heliacal set, the star sets roughly after the beginning of the astronomical night.

Largest Sun separation Happens when the angular separation between the Sun and the celestial object are maximum. In most cases, this is equivalent to say that the Equatorial longitudes of the Sun and the object differ by 180 degrees, so the Sun is in opposition to the object. When an object is at its maximum possible angular separation from the Sun (no matter if it is a planet or a star), it culminates roughly at midnight, and on the darkest possible area of the Sky at that declination. Hence, that is the 'best' night to observe a particular object.

Nights with source above horizon The program computes the range of dates when the celestial object is above the horizon at least during one moment of the night. By 'night', the program considers the time span when the Sun altitude is below that of the twilight (which can be set by the user; see above). When the objects are fixed on the sky (or are exterior planets), the range of observable epochs for the current year can have two possible forms: either a range from one date to another (e.g., 20 Jan to 15 Sep) or in two steps (from 1 Jan to a given date and from another date to 31 Dec). In the first case, the first date (20 Jan in our example) shall be close to the so-called 'Heliacal rise of a star' and the second date (15 Sep in our example) shall be close to the 'Heliacal set'. In the second case (e.g., a range in the form 1 Jan to 20 May and 21 Sep to 31 Dec), the first date (20 May in our example) would be close to the Heliacal set and the second one (21 Sep in our example) to the Heliacal rise. More exact equations to estimate the Heliacal rise/set of stars and planets (which will not depend on the mere input of a twilight Sun elevation by the user) will be implemented in future versions of this plugin.

Full Moon When the Moon is selected, the program can compute the exact closest dates of the Moon's opposition to the Sun.

Author

This plugin has been contributed by Ivan Marti-Vidal (Onsala Space Observatory)[5] with some advice by Alexander Wolf and Georg Zotti.

---

[5]`mailto:i.martividal@gmail.com`

15. Scripting

15.1 Introduction

The development of a powerful scripting system has been continuing for a number of years now and can now be called operational. The use of a script was recognised as a perfect way of arranging a display of a sequence of astronomical events from the earliest versions of Stellarium and a simple system called *Stratoscript* was implemented. The scripting facility is Stellarium's version of a *Presentation*, a feature that may be used to run an astronomical or other presentation for instruction or entertainment from within the Stellarium program. The original *Stratoscript* was quite limited in what it could do so a new Stellarium Scripting System has been developed.

Since version 0.10.1, Stellarium has included a scripting feature based on the Qt Scripting Engine[1]. This makes it possible to write small programs within Stellarium to produce automatic presentations, set up custom configurations, and to automate repetitive tasks.

As of version 0.14.0 a new scripting engine has reached a level where it has all required features for usage, however new commands may be added from time to time. Since version 0.14.0 support of scripts for the *Stratoscript* engine has been discontinued.

The programming language ECMAscript[2] (also known as JavaScript) gives users access to all basic ECMAScript language features such as flow control, variables, string manipulation and so on.

Interaction with Stellarium-specific features is done via a collection of objects which represent components of Stellarium itself. The various modules of Stellarium, and also activated plugins, can be called in scripts to calculate, move the scene, switch on and off display of objects, etc. You can write text output into text files with the output() command. You can call all public slots which are documented in the scripting API documentation[3].

---

[1] http://doc.qt.io/qt-5/qtscript-index.html
[2] https://en.wikipedia.org/wiki/ECMAScript
[3] http://www.stellarium.org/doc/0.15.0/scripting.html

15.2 Script Console

It is possible to open, edit run and save scripts using the script console window. To toggle the script console, press F12. The script console also provides an output window in which script debugging output is visible.

15.3 Includes

Stellarium provides a mechanism for splitting scripts into different files. Typical functions or lists of variables can be stored in separate .inc files and used within other scripts through the **include**() command:

```
include("common_objects.inc");
```

15.4 Minimal Scripts

This script prints "Hello Universe" in the Script Console log window and into log.txt.

```
core.debug("Hello Universe");
```

This script prints "Hello Universe" in the Script Console output window and output.txt.

```
core.output("Hello Universe");
```

The file output.txt will be rewritten on each run of Stellarium. In case you need to save a copy of the current output file to another file, call

```
core.saveOutputAs("myImportantData.txt");
core.resetOutput();
```

This script uses the LabelMgr module to display "Hello Universe" in red, fontsize 20, on the screen for 3 seconds.

```
var label=LabelMgr.labelScreen("Hello Universe", 200, 200,
                                   true, 20, "#ff0000");
core.wait(3);
LabelMgr.deleteLabel(label);
```

15.5 Example: Retrograde motion of Mars

A good way begin writing of scripts: set yourself a specific goal and try to achieve it with the help of few simple steps. Any complex script can be split into simple parts or tasks, which may solve any newbie problems in scripting.

Let me explain it with examples.

Imagine that you have set a goal to make a demonstration of a very beautiful, but longish phenomenon — the retrograde motion of the planet Mars (Fig. 15.1).

15.5.1 Script header...

Any "complex" script should contain a few lines in the first part of the file, which contains important data for humans — the name of the script and its description — and some rules for Stellarium.

```
//
// Name: Retrograde motion of Mars
// Author: John Doe
```

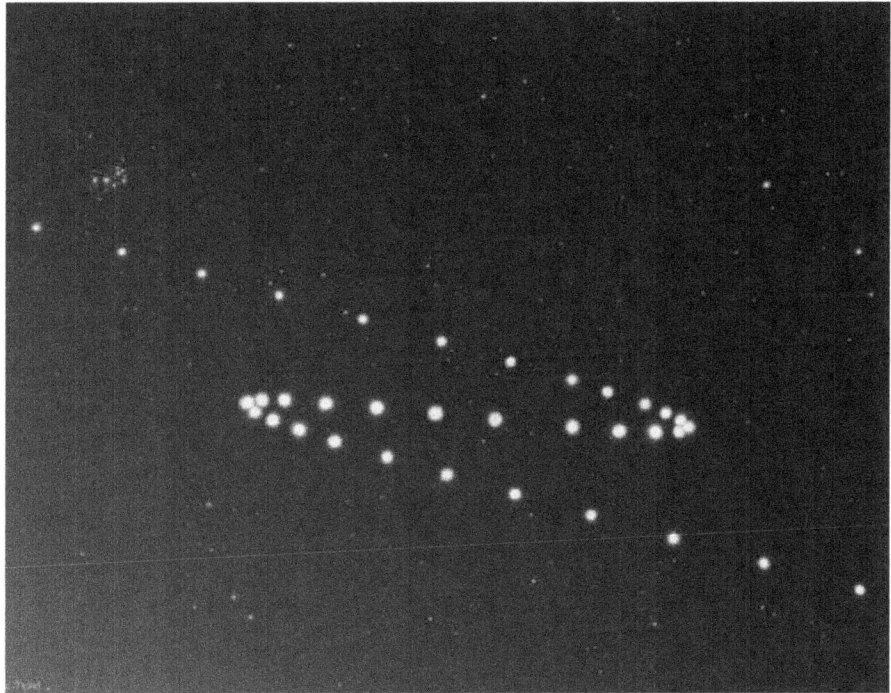

Figure 15.1: Retrograde motion of Mars in 2005. (Credit & Copyright: Tunc Tezel — APOD: 2006 April 22 – Z is for Mars.)

```
// License: Public Domain
// Version: 1.0
// Description: A demo of retrograde motion of Mars.
//
```

15.5.2 A body of script...

At the first stage of writing of the script for a demo of retrograde motion of Mars we should set some limits for our demo. For example we want to see motion of Mars every day during 250 days since October 1st, 2009. Choosing a value of field of view and of the coordinates of the center of the screen should be done at the this stage also.

Let's add few lines of code into the script after the header and run it:

```
core.setDate("2009-10-01T10:00:00");
core.moveToRaDec("08h44m41s", "+18d09m13s",1);
StelMovementMgr.zoomTo(40, 1);
for (i=0; i<250; i++)
{
      core.setDate("+ 1 days");
      core.wait(0.2);
}
```

OK, Stellarium is doing something, but what exactly is it doing? The ground and atmosphere is enabled and any motion of Mars is invisible. Let's add an another few lines into the script (hiding the landscape and atmosphere) after setting date and time:

```
LandscapeMgr.setFlagLandscape(false);
LandscapeMgr.setFlagAtmosphere(false);
```

The whole sky is moving now — let's lock it! Add this line after previous lines:

```
StelMovementMgr.setFlagLockEquPos(true);
```

It looks better now, but what about cardinal points, elements of GUI and some "glitch of movement"? Let's change the script:

```
core.setDate("2009-10-01T10:00:00");
LandscapeMgr.setFlagCardinalsPoints(false);
LandscapeMgr.setFlagLandscape(false);
LandscapeMgr.setFlagAtmosphere(false);
core.setGuiVisible(false);
core.moveToRaDec("08h44m41s", "+18d09m13s",1);
StelMovementMgr.setFlagLockEquPos(true);
StelMovementMgr.zoomTo(40, 1);
core.wait(2);
for (i=0; i<250; i++)
{
        core.setDate("+ 1 days");
        core.wait(0.2);
}
core.setGuiVisible(true);
```

It's better, but let's draw the "path" of Mars! Add those line before loop:

```
core.selectObjectByName("Mars", false);
SolarSystem.setFlagIsolatedTrails(true);
SolarSystem.setFlagTrails(true);
```

Hmm... let's add a few strings with info for users (insert those lines after the header):

```
var color = "#ff9900";
var info = LabelMgr.labelScreen("A motion of Mars", 20, 20,
          false, 24, color);
var apx = LabelMgr.labelScreen("Setup best viewing angle, FOV
          and date/time.", 20, 50, false, 18, color);
LabelMgr.setLabelShow(info, true);
LabelMgr.setLabelShow(apx, true);
core.wait(2);
LabelMgr.setLabelShow(apx, false);
```

Let's add some improvements to display info for users — change in the loop:

```
var label = LabelMgr.labelObject("  Normal motion, West to
          East", "Mars", true, 16, color, "SE");
for (i=0; i<250; i++)
{
        core.setDate("+ 1 days");
        if ((i % 10) == 0)
        {
                var strDate = "Day " + i;
                LabelMgr.setLabelShow(apx, false);
                var apx = LabelMgr.labelScreen(strDate, 20,
                          50, false, 16, color);
                LabelMgr.setLabelShow(apx, true);
```

```
        }
        if (i == 75)
        {
                LabelMgr.deleteLabel(label);
                label = LabelMgr.labelObject(" Retrograde or
                        opposite motion begins", "Mars",
                        true, 16, color, "SE");
                core.wait(2);
                LabelMgr.deleteLabel(label);
                label = LabelMgr.labelObject(" Retrograde
                        motion", "Mars", true, 16, color,
                        "SE");
        }
        if (i == 160)
        {
                LabelMgr.deleteLabel(label);
                label = LabelMgr.labelObject(" Normal motion
                        returns", "Mars", true, 16, color,
                        "SE");
                core.wait(2);
                LabelMgr.deleteLabel(label);
                label = LabelMgr.labelObject(" Normal motion",
                        "Mars", true, 16, color, "SE");
        }
        core.wait(0.2);
}
```

15.6 More Examples

The best source of examples is the `scripts` sub-directory of the main Stellarium source tree. This directory contains a sub-directory called `tests` which are not installed with Stellarium, but are nonetheless useful sources of example code for various scripting features[4].

---

[4]The directory can be browsed online at `http://bazaar.launchpad.net/~stellarium/ stellarium/trunk/files/head:/scripts/`. Script files end in `.ssc` and include files (which are not runnable by themselves) in `.inc`. Download links are to the right.

IV Practical Astronomy

16   Astronomical Concepts 169

17   Astronomical Phenomena 185

18   A Little Sky Guide 201

19   Exercises 211

16. Astronomical Concepts

Barry Gerdes, with additions by Georg Zotti

This section includes some general notes on astronomy in an effort to outline some concepts that are helpful to understand features of Stellarium. Material here is only an overview, and the reader is encouraged to get hold of a couple of good books on the subject. A good place to start is a compact guide and ephemeris such as the *National Audubon Society Field Guide to the Night Sky*[1]. Also recommended is a more complete textbook such as *Universe*. There are also some nice resources on the net, like the *Wikibooks Astronomy book*[2].

16.1 The Celestial Sphere

The *Celestial Sphere* is a concept which helps us think about the positions of objects in the sky. Looking up at the sky, you might imagine that it is a huge dome or top half of a sphere, and the stars are points of light on that sphere. Visualising the sky in such a manner, it appears that the sphere moves, taking all the stars with it — it seems to rotate. Watching the movement of the stars we can see that they seem to rotate around a static point about once a day. Stellarium is the perfect tool to demonstrate this!

1. Open the location dialog (F6). Set the location to be somewhere in mid-Northern latitudes. (Just click on the map to select a location, or fine-tune with the settings.) The United Kingdom is an ideal location for this demonstration.
2. Turn off atmospheric rendering A and ensure cardinal points are turned on (Q). This will keep the sky dark so the Sun doesn't prevent us from seeing the motion of the stars when it is above the horizon.
3. Pan round to point North, and make sure the field of view is about 90°.
4. Pan up so the 'N' cardinal point on the horizon is at the bottom of the screen.

---

[1]http://www.amazon.com/National-Audubon-Society-Field-Series/dp/0679408525
[2]http://en.wikibooks.org/wiki/Subject:Astronomy

5. Now increase the time rate. Press K , L , L , L , L – this should set the time rate so the stars can be seen to rotate around a point in the sky about once every ten seconds. If you watch Stellarium's clock you'll see this is the time it takes for one day to pass at this accelerated rate.

The point which the stars appear to move around is one of the *Celestial Poles*.

The apparent movement of the stars is due to the rotation of the Earth. Our location as the observer on the surface of the Earth affects how we perceive the motion of the stars. To an observer standing at Earth's North Pole, the stars all seem to rotate around the *zenith* (the point directly upward). As the observer moves South towards the equator, the location of the celestial pole moves down towards the horizon. At the Earth's equator, the North celestial pole appears to be on the Northern horizon.

Similarly, observers in the Southern hemisphere see the Southern celestial pole at the zenith when they are at the South pole, and it moves to the horizon as the observer travels towards the equator.

1. Leave time moving on nice and fast, and open the configuration window. Go to the location tab and click on the map right at the top – i.e., set your location to the North pole. See how the stars rotate parallel to the horizon, around a point right at the top of the screen. With the field of view set to 90° and the horizon at the bottom of the screen, the top of the screen is the zenith.

2. Now click on the map again, this time a little further South. You should see the positions of the stars jump, and the centre of rotation has moved a little further down the screen.

3. Click on the map even further towards and equator. You should see the centre of rotation having moved down again.

To help with the visualisation of the celestial sphere, turn on the equatorial grid by clicking the button on the main tool-bar or pressing the E key. Now you can see grid lines drawn on the sky. These lines are like lines of longitude and latitude on the Earth, but drawn for the celestial sphere.

The *Celestial Equator* is the line around the celestial sphere that is half way between the celestial poles – just as the Earth's equator is the line half way between the Earth's poles.

16.2 Coordinate Systems

16.2.1 Altitude/Azimuth Coordinates

The *Altitude/Azimuth* coordinate system (also called *Horizontal Coordinate System*) can be used to describe a direction of view (the *azimuth* angle) and an angular height in the sky (the *altitude* angle). The azimuth angle is measured clockwise round from due North[3]. Hence North itself is 0°, East 90°, Southwest is 225° and so on. The altitude angle is measured up from the *mathematical horizon*, which is just halfway between "straight up" and "straight down", without regard to the landscape. Looking directly up (at the *zenith*) would be 90°, half way between the zenith and the horizon is 45° and so on. The point opposite the zenith is called the *nadir*.

The Altitude/Azimuth coordinate system is attractive in that it is intuitive – most people are familiar with azimuth angles from bearings in the context of navigation, and the altitude angle is something most people can visualise pretty easily.

However, the altitude/azimuth coordinate system is not suitable for describing the general position of stars and other objects in the sky – the altitude and azimuth values for a celestial object change with time and the location of the observer.

---

[3]In some textbooks azimuth is counted from south. There is no global authority to decide upon this issue, just be aware of this when you compare numbers with other sources.

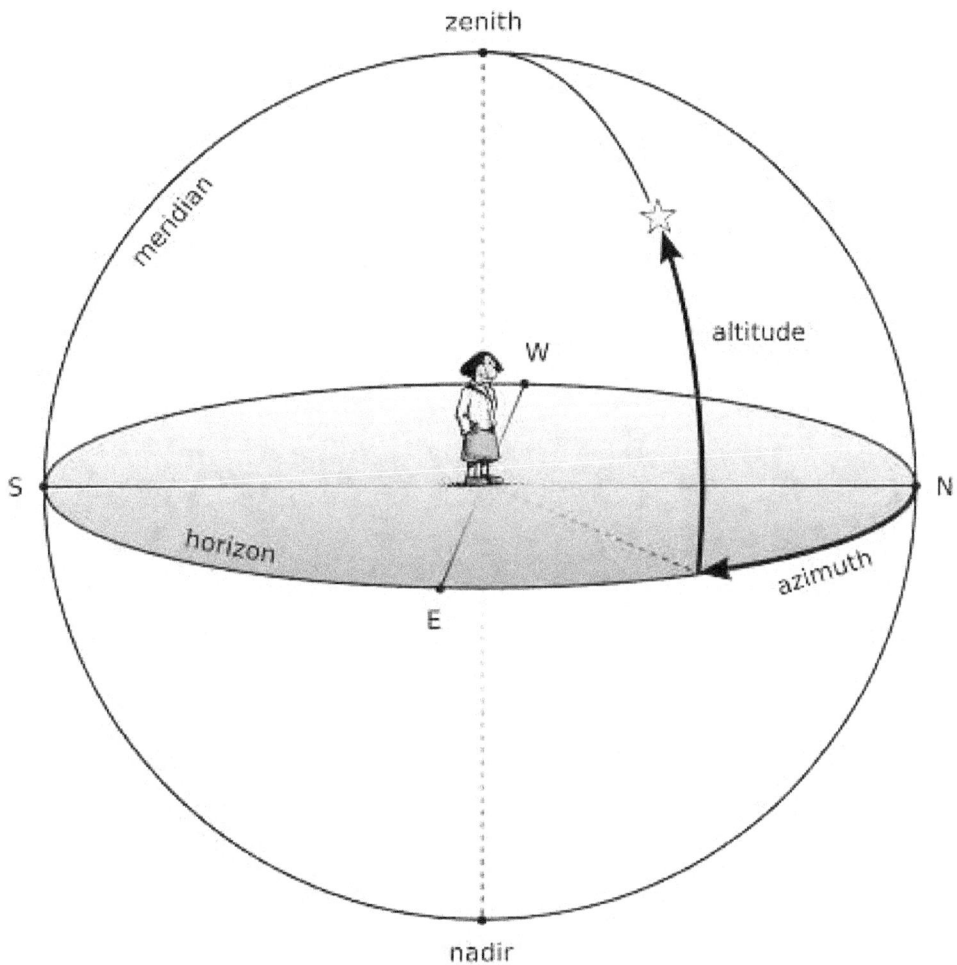

Figure 16.1: Altitude/Azimuth (Horizontal) Coordinate System

Stellarium can draw grid lines for altitude/azimuth coordinates. Use the [⊕] button on the main tool-bar to activate this grid, or press the [Z] key.

In addition, the *cardinal points* can be highlighted using the [✛] button or [Q] key.

There are a few great circles with special names which Stellarium can draw (see section 4.4.3).

Meridian This is the vertical line which runs from the North point towards the zenith and further to the South point.

(Mathematical) Horizon This is the line exactly 90° away from the zenith.

First Vertical This is the vertical line which runs from the East point towards the zenith and further to the West point.

16.2.2 Right Ascension/Declination Coordinates

Like the Altitude/Azimuth system, the *Right Ascension/Declination* (RA/Dec) Coordinate System (or *Equatorial Coordinate System*) uses two angles to describe positions in the sky. These angles

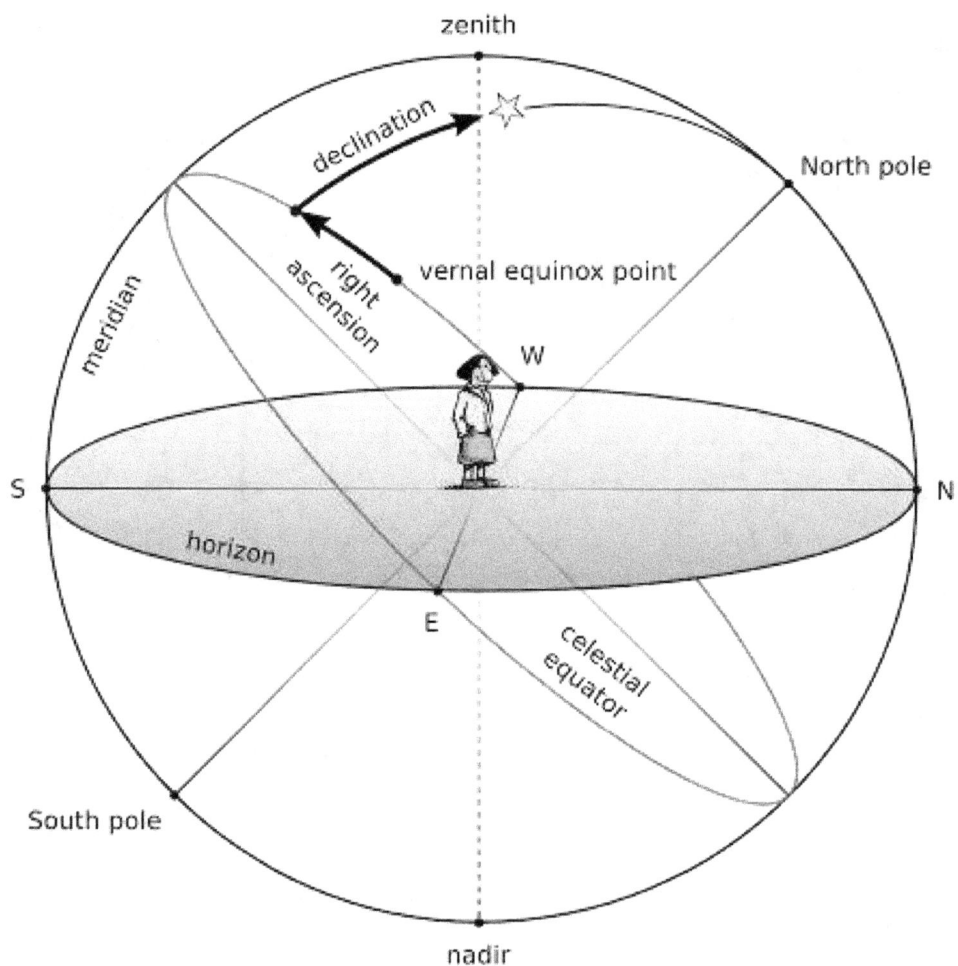

Figure 16.2: Equatorial Coordinates

are measured from standard points on the celestial sphere. *Right ascension* α and *declination* δ are to the celestial sphere what longitude and latitude are to terrestrial map makers.

The Northern celestial pole has a declination of $\delta = 90°$, the celestial equator has a declination of $\delta = 0°$, and the Southern celestial pole has a declination of $\delta = -90°$.

Right ascension is measured as an angle round from a point in the sky known as the *First Point of Aries*, in the same way that longitude is measured around the Earth from Greenwich. Figure 16.2 illustrates RA/Dec coordinates. The angle α is usually expressed as time with minute and seconds, with 15° equalling one hour.

Unlike Altitude/Azimuth coordinates, RA/Dec coordinates of a star do not change if the observer changes latitude, and do not change noticeably over the course of the day due to the rotation of the Earth. RA/Dec coordinates are generally used nowadays in star catalogues such as the Hipparcos catalogue.

However, the story is complicated a little by precession (section 16.4) and parallax (section 16.5). Precession causes a slow drift of the coordinates almost parallel to the ecliptic, and therefore star catalogues always have to specify their *equinox* of validity. Current catalogs and atlases use coordinates for J2000.0.

Stellarium can draw grid lines for Equatorial coordinates. Use the button ⊕ on the main tool-bar to activate this grid, or press the ⌷ E ⌷ key to draw the equatorial grid for the simulation time. The Markings dialog (4.4.3) allows you to set also the grid for J2000.0 standard coordinates.

In case you are observing from another celestial object, the equatorial coordinates use a system similar to the one referring to the earth-based coordinates, but parallel to the planet's rotational axis.

There are again a few great circles with special names which Stellarium can draw in addition, both for simulation time and for J2000.0 (see section 4.4.3).

Celestial Equator the line directly above the earth's (or more generally, the observer's planet's) equator.

Colures These are lines similar to meridian and first vertical in the azimuthal system. The *Equinoctial Colure* runs from the North Celestial Pole NCP through the First Point of Aries ♈, South Celstial Pole SCP and First Point of Libra ♎ while the *Solstitial Colure* runs from the NCP through First Point of Cancer ♋, SCP and First Point of Capricorn ♑.

16.2.3 Ecliptical Coordinates

The earth's orbit around the sun, i.e., the *ecliptic*, defines the "equatorial line" of this coordinate system, which is traditionally used when computing the coordinates for planets.

The zero point of *ecliptical longitude* λ is the same as for equatorial coordinates, and the *ecliptical latitude* β is counted positive towards the *Northern Ecliptical Pole* NEP in the constellation of Draco.

Moon and sun (and to a much lesser extent, the other planets) pull on the equatorial bulge and try to put Earth's axis normal to its orbital plane. Earth acts like a spinning top and evades this pull in a sideways motion, so that earth's axis seems to describe a small circle over a period of almost 26.000 years (see section 16.4).

In addition, ecliptic obliquity against the equatorial coordinates, which mirrors the earth's axial tilt, slowly changes.

Therefore, also for ecliptical coordinates is is required to specify which date the coordinates refer to. Stellarium can draw two grids for Ecliptical coordinates. Use the ⌷ , ⌷ key to draw the ecliptic for the simulation time. The Markings dialog (4.4.3) allows you to show also a line for epoch J2000.0 and grids for the ecliptical coordinates for current epoch and epoch J2000.0. You can assign your own shortcut keys (section 4.7.1) if you frequently operate with these coordinates.

Since version 0.14.0 Stellarium can very accurately show the motions between the coordinate systems [67], and it is quite interesting to follow these motions for several millennia. To support such demonstrations, Stellarium can also draw the *precession circles* between celestial and ecliptical poles (activate them in the Markings dialog (4.4.3). If you observe long enough, you will see that these circles vary in size, reflecting the changes in ecliptic obliquity.

Many of the minor bodies are best observed around the times of their opposition. Stellarium can display a great circle in the ecliptical coordinates which runs through the ecliptic poles and through the sun, thereby allowing to estimate opposition and conjunction. Activate display of this *Opposition/Conjunction Line* in the Markings dialog (Labeled "O./C. longitude"; 4.4.3).

It is interesting to note that star catalogs before TYCHO BRAHE's (1546–1601), most notably the one in PTOLEMY's *Almagest*, used Ecliptical coordinates. The reason is simple: It was known since HIPPARCH that stellar coordinates slowly move along the ecliptic through *precession*, and the correction to coordinates of a date of interest was a simple addition of a linear correction to the ecliptical longitude in the catalog. Changes of ecliptic obliquity was discovered much later.

16.2.4 **Galactic Coordinates**

The Milky Way appears to run along a great circle over the sky, mirroring the fact that the sun is a star in it. Coordinates for non-stellar objects which belong to the Milky Way like *pulsars* or *planetary nebulae* are often mapped in *Galactic Coordinates*, where *galactic longitude l* and *galactic latitude b* are usually given in decimal degrees. Here, the zero point of galactic longitudes lies in the Galactic Center.

Stellarium can also draw a galactic grid and the galactic equator by activating the respective options in the Markings dialog (see section 4.4.3). You can assign a keyboard shortcut if you frequently use these coordinates (see 4.7.1).

16.3 **Units**

16.3.1 **Distance**

As DOUGLAS ADAMS pointed out in the Hitchhiker's Guide to the Galaxy[2],

> Space [...] is big. Really big. You just won't believe how vastly, hugely, mind-bogglingly big it is. I mean, you may think it's a long way down the road to the chemist, but that's just peanuts to space.[p.76]

Astronomers use a variety of units for distance that make sense in the context of the mind-boggling vastness of space.

Astronomical Unit (AU) This is the mean Earth-Sun distance. Roughly 150 million kilometres (1.49598×10^8 km). The AU is used mainly when discussing the solar system – for example the distance of various planets from the Sun.

Light year (LY) A light year is not, as some people believe, a measure of time. It is the distance that light travels in a year. The speed of light being approximately 300,000 kilometres per second means a light year is a very large distance indeed, working out at about 9.5 trillion kilometres (9.46073×10^{12} km). Light years are most frequently used when describing the distance of stars and galaxies or the sizes of large-scale objects like galaxies, nebulae etc.

Parsec (pc) A parsec is defined as the distance of an object that has an annual parallax of 1 second of arc. This equates to 3.26156 light years (3.08568×10^{13} km). Parsecs (and derivatives: kiloparsec kpc, megaparsec Mpc) are most frequently used when describing the distance of stars or the sizes of large-scale objects like galaxies, nebulae etc.

16.3.2 **Time**

The length of a day is defined as the amount of time that it takes for the Sun to travel from the highest point in the sky at mid-day to the next high-point on the next day. In astronomy this is called a *solar day*. The apparent motion of the Sun is caused by the rotation of the Earth. However, in this time, the Earth not only spins, it also moves slightly round its orbit. Thus in one solar day the Earth does not spin exactly 360° on its axis. Another way to measure day length is to consider how long it takes for the Earth to rotate exactly 360°. This is known as one *sidereal day*.

Figure 16.3 illustrates the motion of the Earth as seen looking down on the Earth orbiting the Sun. The red triangle on the Earth represents the location of an observer. The figure shows the Earth at four times:

1. The Sun is directly overhead - it is mid-day.
2. Twelve hours have passed since 1. The Earth has rotated round and the observer is on the opposite side of the Earth from the Sun. It is mid-night. The Earth has also moved round in its orbit a little.
3. The Earth has rotated exactly 360°. Exactly one sidereal day has passed since 1.

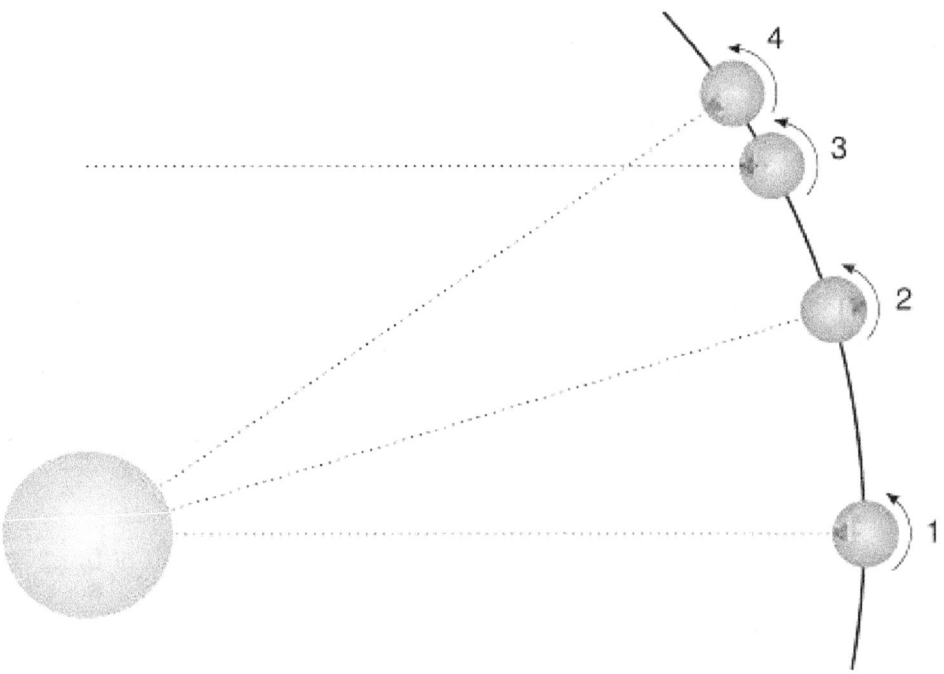

Figure 16.3: Sidereal day

4. It is mid-day again – exactly one solar day since 1. Note that the Earth has rotated more than 360° since 1.

It should be noted that in figure 16.3 the sizes of the Sun and Earth and not to scale. More importantly, the distance the Earth moves around its orbit is much exaggerated. The Earth takes a year to travel round the Sun – $365\frac{1}{4}$ solar days. The length of a sidereal day is about 23 hours, 56 minutes and 4 seconds.

Sidereal Time

It takes exactly one sidereal day for the celestial sphere to make one revolution in the sky. Astronomers find *sidereal time* useful when observing. This is the Right Ascension which is currently passing the meridian line. When visiting observatories, look out for doctored alarm clocks that have been set to run in sidereal time!

16.3.3 Julian Day Number

In the 19th century, astronomer JOHN HERSCHEL introduced the use of *Julian Day* numbers (invented around the time of the Gregorian calendar reform). This is a simple continuous day count starting on January 1, -4712 (4713 BC). There are no years, months etc., and the integral day number switches at noon, so during a single night of observation (in Europe) the date never changes.

The fractional part of the number is just the fraction of day that has elapsed since last noon. Given that a day has 86400 seconds, we should give a JD with 5 decimal places to capture the nearest second.

This causes a problem for modern computers: even a "double precision float" can keep only about 13 decimal places. More than 2.4 million days have passed, so that e.g. January 1, 2000, 12:00UT is 2451545.0, which is an accurately storable number with 7 decimal places, but 12:34:56UT is computed as 2451545.02426. A more accurate result would yield 2451545.024259259259... So, for a field where sub-second accuracy became crucial like spacecraft operations, the *Modified Julian Date* (MJD) has been introduced. It is simply

$$MJD = JD - 2400000.5. \tag{16.1}$$

This means, days start at midnight, and the (constant, in our era) decimal places of the "big numbers" at the begin of the number have been traded in for more decimal places at the end.

Don't put your expectations too high when you see MJD displayed (section 4.1): Stellarium uses a double-precision floating point number for JD for internal timekeeping, and Stellarium's display of MJD is simply computed from it. So you cannot set temporal increments smaller than a second, and it hardly would make sense to expect more accuracy from the simulation algorithms.

Delta T

Until around 1900, the earth's rotation was regarded as perfect standard of time. There were 86400 seconds per mean solar day, and the accuracy of reproducing time with mechanical clocks only in this time started to become as good as the earth's rotation itself.

Astronomers who computed solar eclipses reported in texts from antiquity wondered about a required time shift which they originally attributed to a yet-unknown "secular acceleration of the lunar motion". However, it turned out that indeed the gravitational effect of the moon which causes the tides also has effects on earth's rotation: the tides slowly break earth's rotational speed. The energy is also transferred to the moon, and the acceleration leads to the moon slowly moving away from the earth[4].

This led to the introduction of a time named Ephemeris Time (ET) with progresses in the speed of the second in the year 1900, to be used for positional computation in our solar system, in addition to Greenwich Mean Time (GMT), from which all zone times and "civil" clock times were derived.

The introduction of Atomic Clocks in the middle of the 20th century led to a redefinition of the (temporal) second, which has been de-coupled from earth's rotation. This time, the International Atomic Time TAI, is the basis for Terrestrial Time TT which can be considered as constantly progressing at constant speed[5], and is used for computation of the planetary positions.

Still, people living on earth prefer to have the mean solar noon governing the run of day and night. Therefore all forms of civil time is linked to Coordinated Universal Time UTC. Seconds in UTC and TAI are of equal length. The slow and irregular divergence between TAI and UTC is observed by a few standardisation institutes. When necessary, a leap second can be introduced to the UTC to bring the earth's rotation back in sync so that the Mean Sun again culminates at noon.

The difference $\Delta T = TT - UT$ (or "Delta T") describes the temporal offset which amounts already to more than a minute in the 21st century. There have been many attempts to properly model ΔT, and Stellarium offers several models you can choose from in the configuration dialog (see section 4.3.3). The default, "Espenak and Meeus (2006)", is a widely accepted standard. But if you are a researcher and want to experiment with alternative models, you will hopefully like this feature. you can even specify your own data for a, b, c, y and the secular term for lunar acceleration n

---

[4]No need to worry, the moon recedes from the earth only a few centimeters per year as measured with the laser reflectors left by the Apollo astronauts in the 1970s. In a very far future, however, there will only be annular solar eclipses as a consequence!

[5]We don't discuss relativity here. The advanced reader is referred to the presentation in the Wikipedia, https://en.wikipedia.org/wiki/Delta_T.

(actually $\dot{n} = dn/dt$ in units of arcseconds/century^2) if you can model ΔT according to the formula

$$\Delta T = a + b \cdot u + c \cdot u^2 \quad \text{where} \tag{16.2}$$
$$u = \frac{\text{year} - y}{100} \tag{16.3}$$

List of equations of ΔT in Stellarium

The following list describes sources and a few details about the models for ΔT implemented in Stellarium.

Without correction. Correction is disabled. Use only if you know what you are doing!

Schoch (1931). This historical formula was obtained by C. Schoch [55] and was used by G. Henriksson in his article "Einstein's Theory of Relativity Confirmed by Ancient Solar Eclipses" [23]. See for more in [48]. $\dot{n} = -29.68''/\text{cy}^2$.

Clemence (1948). This empirical equation was published by G. M. Clemence in the article "On the system of astronomical constants" [15]. Valid range of usage: between years 1681 and 1900. $\dot{n} = -22.44''/\text{cy}^2$.

IAU (1952). This formula is based on a study of post-1650 observations of the Sun, the Moon and the planets by Spencer Jones [58] and used by Jean Meeus in his "Astronomical Formulae for Calculators". It was also adopted in the PC program **SunTracker Pro**. Valid range of usage: between years 1681 and 1936. $\dot{n} = -22.44''/\text{cy}^2$.

Astronomical Ephemeris (1960). This is a slightly modified version of the IAU [58] formula which was adopted in the "Astronomical Ephemeris" [21] and in the "Canon of Solar Eclipses" by Mucke & Meeus [45]. Valid range of usage: between years -500 and 2000. $\dot{n} = -22.44''/\text{cy}^2$.

Tuckerman (1962, 1964) & Goldstine (1973). The tables of Tuckerman [63, 64] list the positions of the Sun, the Moon and the planets at 5- and 10-day intervals from 601 BCE to 1649 CE. The same relation was also implicitly adopted in the syzygy tables of Goldstine [22]. Valid range of usage: between years -600 and 1649.

Muller & Stephenson (1975). This equation was published by P. M. Muller and F. R. Stephenson in the article "The accelerations of the earth and moon from early astronomical observations" [46]. Valid range of usage: between years -1375 and 1975. $\dot{n} = -37.5''/\text{cy}^2$.

Stephenson (1978). This equation was published by F. R. Stephenson in the article "Pre-Telescopic Astronomical Observations" [59]. $\dot{n} = -30.0''/\text{cy}^2$.

Schmadel & Zech (1979). This 12th-order polynomial equation (outdated and superseded by Schmadel & Zech (1988)) was published by L. D. Schmadel and G. Zech in the article "Polynomial approximations for the correction delta T E.T.-U.T. in the period 1800-1975" [53] as fit through data published by Brouwer [10]. Valid range of usage: between years 1800 and 1975, with meaningless values outside this range. $\dot{n} = -23.8946''/\text{cy}^2$.

Morrison & Stephenson (1982). This algorithm [42] was adopted in P. Bretagnon & L. Simon's "Planetary Programs and Tables from -4000 to +2800" [9] and in the PC planetarium program **RedShift**. Valid range of usage: between years -4000 and 2800. $\dot{n} = -26.0''/\text{cy}^2$.

Stephenson & Morrison (1984). This formula was published by F. R. Stephenson and L. V. Morrison in the article "Long-term changes in the rotation of the earth - 700 B.C. to A.D. 1980" [60]. Valid range of usage: between years -391 and 1600. $\dot{n} = -26.0''/\text{cy}^2$.

Stephenson & Houlden (1986). This algorithm [25] is used in the PC planetarium program **Guide 7**. Valid range of usage: between years -600 and 1600. $\dot{n} = -26.0''/\text{cy}^2$.

Espenak (1987, 1989). This algorithm was given by F. Espenak in his "Fifty Year Canon of Solar Eclipses: 1986-2035" [18] and in his "Fifty Year Canon of Lunar Eclipses: 1986-2035" [19]. Valid range of usage: between years 1950 and 2100.

Borkowski (1988). This formula was obtained by K.M. Borkowski [6] from an analysis of 31 solar eclipse records dating between 2137 BCE and 1715 CE. Valid range of usage: between years -2136 and 1715. $\dot{n} = -23.895''/\text{cy}^2$.

Schmadel & Zech (1988). This 12th-order polynomial equation was published by L. D. Schmadel and G. Zech in the article "Empirical Transformations from U.T. to E.T. for the Period 1800-1988" [54] as data fit through values given by Stephenson & Morrison (1984). Valid range of usage: between years 1800 and 1988, with a mean error of less than one second, max. error 1.9s, and meaningless values outside this range. $\dot{n} = -26.0''/\text{cy}^2$.

Chapront-Touze & Chapront (1991). This formula was adopted by M. Chapront-Touze & J. Chapront in the shortened version of the ELP 2000-85 lunar theory in their "Lunar Tables and Programs from 4000 B.C. to A.D. 8000" [14]. The relations are based on those of Stephenson & Morrison (1984), but slightly modified to make them compatible with the tidal acceleration parameter of $\dot{n} = -23.8946''/\text{cy}^2$ adopted in the ELP 2000-85 lunar theory.

Stephenson & Morrison (1995). This equation was published by F. R. Stephenson and L. V. Morrison in the article "Long-Term Fluctuations in the Earth's Rotation: 700 BC to AD 1990" [61]. Valid range of usage: between years -700 and 1600. $\dot{n} = -26.0''/\text{cy}^2$.

Stephenson (1997). F. R. Stephenson published this formula in his book "Historical Eclipses and Earth's Rotation" [62]. Valid range of usage: between years -500 and 1600. $\dot{n} = -26.0''/\text{cy}^2$.

Meeus (1998) (with Chapront, Chapront-Touze & Francou (1997)). From J. Meeus, "Astronomical Algorithms" [37], and widely used. Table for 1620..2000, and includes a variant of Chapront, Chapront-Touze & Francou (1997) for dates outside 1620..2000. Valid range of usage: between years -400 and 2150. $\dot{n} = -25.7376''/\text{cy}^2$.

JPL Horizons. The JPL Solar System Dynamics Group of the NASA Jet Propulsion Laboratory use this formula in their interactive website JPL Horizons[6]. Valid range of usage: between years -2999 and 1620, with zero values outside this range. $\dot{n} = -25.7376''/\text{cy}^2$.

Meeus & Simons (2000). This polynome was published by J. Meeus and L. Simons in article "Polynomial approximations to Delta T, 1620-2000 AD" [34]. Valid range of usage: between years 1620 and 2000, with zero values outside this range. $\dot{n} = -25.7376''/\text{cy}^2$.

Montenbruck & Pfleger (2000). The fourth edition of O. Montenbruck & T. Pfleger's "Astronomy on the Personal Computer" [41] provides simple 3rd-order polynomial data fits for the recent past. Valid range of usage: between years 1825 and 2005, with a typical 1-second accuracy and zero values outside this range.

Reingold & Dershowitz (2002, 2007). E. M. Reingold & N. Dershowitz present this polynomial data fit in "Calendrical Calculations" [50] and in their "Calendrical Tabulations" [49]. It is based on Jean Meeus' "Astronomical Algorithms" [35].

Morrison & Stephenson (2004, 2005). This important solution was published by L. V. Morrison and F. R. Stephenson in article "Historical values of the Earth's clock error ΔT and the calculation of eclipses" [43] with addendum [44]. Valid range of usage: between years -1000 and 2000. $\dot{n} = -26.0''/\text{cy}^2$.

Espenak & Meeus (2006). This solution[7] by F. Espenak and J. Meeus, based on Morrison & Stephenson [43] and a polynomial fit through tabulated values for 1600-2000, is used for the NASA Eclipse Web Site[8] and in their "Five Millennium Canon of Solar Eclipses: -1900 to +3000" [20]. This formula is also used in the solar, lunar and planetary ephemeris program **SOLEX**. Valid range of usage: between years -1999 and 3000. $\dot{n} = -25.858''/\text{cy}^2$.

Reijs (2006). From the Length of Day (LOD; as determined by Stephenson & Morrison [43]), Victor Reijs derived a ΔT formula by using a Simplex optimisation with a cosine and square

---

[6]http://ssd.jpl.nasa.gov/?horizons
[7]This solution is used by default.
[8]http://eclipse.gsfc.nasa.gov/eclipse.html

function[9]. This is based on a possible periodicy described by Stephenson [43]. Valid range of usage: between years -1500 and 1100. $\dot{n} = -26.0''/\text{cy}^2$.

Banjevic (2006). This solution by B. Banjevic, based on Stephenson & Morrison (1984) [60], was published in article "Ancient eclipses and dating the fall of Babylon" [4]. Valid range of usage: between years -2020 and 1620, with zero values outside this range. $\dot{n} = -26.0''/\text{cy}^2$.

Islam, Sadiq & Qureshi (2008, 2013). This solution by S. Islam, M. Sadiq and M. S. Qureshi, based on Meeus & Simons [34], was published in article "Error Minimization of Polynomial Approximation of DeltaT" [26] and revisited by Sana Islam in 2013. Valid range of usage: between years 1620 and 2007, with zero values outside this range.

Khalid, Sultana & Zaidi (2014). This polynomial approximation with 0.6 seconds of accuracy by M. Khalid, Mariam Sultana and Faheem Zaidi was published in "Delta T: Polynomial Approximation of Time Period 1620-2013" [27]. Valid range of usage: between years 1620 and 2013, with zero values outside this range.

Custom equation of ΔT. This is a quadratic formula for calculation of ΔT with coefficients defined by the user.

16.3.4 Angles

Astronomers typically use degrees to measure angles. Since many observations require very precise measurement, the degree is subdivided into sixty *minutes of arc* also known as *arc-minutes*. Each minute of arc is further subdivided into sixty *seconds of arc*, or *arc-seconds*. Thus one degree is equal to 3600 seconds of arc. Finer grades of precision are usually expressed using the SI prefixes with arc-seconds, e.g. *milli arc-seconds* (one milli arc-second is one thousandth of an arc-second).

Notation

Degrees are denoted using the $°$ symbol after a number. Minutes of arc are denoted with a $'$, and seconds of arc are denoted using $''$. Angles are frequently given in two formats:

1. DMS format — degrees, minutes and seconds. For example $90°15'12''$. When more precision is required, the seconds component may include a decimal part, for example $90°15'12.432''$.
2. Decimal degrees, for example $90.2533°$

Handy Angles

Being able to estimate angular distance can be very useful when trying to find objects from star maps in the sky. One way to do this with a device called a *crossbow*.

Crossbows are a nice way get an idea of angular distances, but carrying one about is a little cumbersome. A more convenient alternative is to hold up an object such as a pencil at arm's length. If you know the length of the pencil, d, and the distance of it from your eye, D, you can calculate its angular size, θ using this formula:

$$\theta = 2 \cdot \arctan\left(\frac{d}{2 \cdot D}\right) \tag{16.4}$$

Another, more handy (ahem!) method is to use the size of your hand at arm's length:

Tip of little finger About $1°$
Middle three fingers About $4°$
Across the knuckles of the fist About $10°$
Open hand About $18°$

Using you hand in this way is not very precise, but it's close enough to give you some way to translate an idea like "Mars will be $45°$ above the Southeastern horizon at 21:30". Of course,

---

[9]http://www.iol.ie/~geniet/eng/DeltaTeval.htm

there is variation from person to person, but the variation is compensated for somewhat by the fact that people with long arms tend to have larger hands. In exercise 19.2 you will work out your own "handy angles".

16.3.5 The Magnitude Scale

When astronomers talk about magnitude, they are referring to the brightness of an object. How bright an object appears to be depends on how much light it is giving out and how far it is from the observer. Astronomers separate these factors by using two measures: *absolute magnitude* (Mag or *M*) which is a measure of how much light is being given out by an object, and *apparent magnitude* (mag or *m*) which is how bright something appears to be in the sky.

For example, consider two 100 watt lamps, one which is a few meters away, and one which is a kilometre away. Both give out the same amount of light – they have the same absolute magnitude. However the nearby lamp seems much brighter – it has a much greater apparent magnitude. When astronomers talk about magnitude without specifying whether they mean apparent or absolute magnitude, they are usually referring to apparent magnitude.

The magnitude scale has its roots in antiquity. The Greek astronomer HIPPARCHUS defined the brightest stars in the sky to be *first magnitude*, and the dimmest visible to the naked eye to be *sixth magnitude*. In the 19th century British astronomer NORMAN POGSON quantified the scale more precisely, defining it as a logarithmic scale where a magnitude 1 object is 100 times as bright as a magnitude 6 object (a difference of five magnitudes). The zero-point of the modern scale was originally defined as the brightness of the star Vega, however this was re-defined more formally in 1982[**landolt**]. Objects brighter than Vega are given negative magnitudes.

The absolute magnitude of a star is defined as the magnitude a star would appear if it were 10 parsecs from the observer.

Table 16.2 lists several objects that may be seen in the sky, their apparent magnitude and their absolute magnitude where applicable (only stars have an absolute magnitude value. The planets and the Moon don't give out light like a star does – they reflect the light from the Sun).

Object	*m*	*M*
The Sun	-27	4.8
Vega	0.05	0.6
Betelgeuse	0.47	-7.2
Sirius (the brightest star)	-1.5	1.4
Venus (at brightest)	-4.4	—
Full Moon (at brightest)	-12.6	—

Table 16.2: Magnitudes of a few objects

16.3.6 Luminosity

Luminosity is an expression of the total energy radiated by a star. It may be measured in watts, however, astronomers tend to use another expression — *solar luminosities* where an object with twice the Sun's luminosity is considered to have two solar luminosities and so on. Luminosity is related to absolute magnitude.

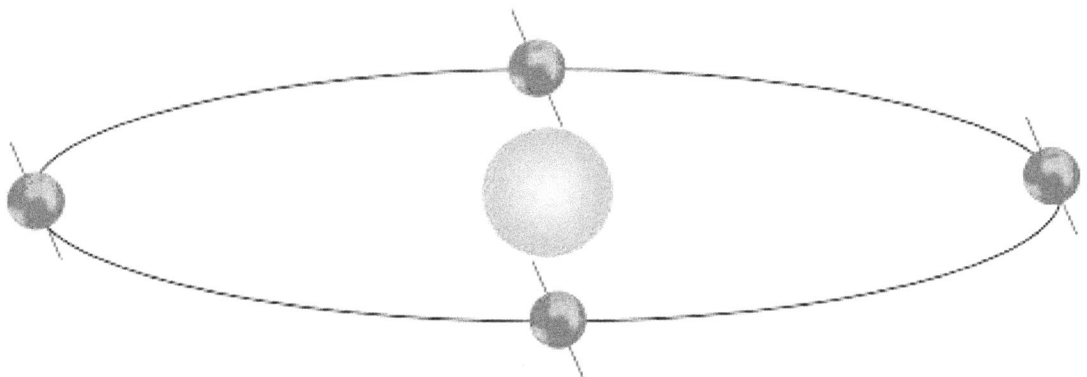

Figure 16.4: Ecliptic obliquity

16.4 Precession

As the Earth orbits the Sun throughout the year, the axis of rotation (the line running through the rotational poles of the Earth) seems to point towards the same position on the celestial sphere, as can be seen in figure 16.4. The angle between the axis of rotation and the perpendicular of the orbital plane is called the *obliquity of the ecliptic*. It is currently about $23°27'$ and is the angle between equatorial coordinates (16.2.2) and ecliptical coordinates (16.2.3).

Observed over very long periods of time the direction the axis of rotation points to does actually change. The angle between the axis of rotation and the orbital plane stays fairly constant, but the direction the axis points — the position of the celestial pole — transcribes a figure similar to a circle on the stars in the celestial sphere. The motion is similar to the way in which a gyroscope slowly twists, as figure 16.5 illustrates. This process is called *precession*. The circles can be shown in Stellarium: From the View menu (F4), tab "Markings", switch on "Precession Circles" (4.4.3).

Precession is a slow process. The axis of rotation twists through a full $360°$ about once every 26,000 years. However, over these long times other gravitational perturbations ("planetary precession") play a role, and what may be thought of as rigid "precession circle" can actually only show the instantaneous (current) state. Over millennia the circle slightly varies.

Precession has some important implications:

1. RA/Dec coordinates change over time, albeit slowly. Measurements of the positions of stars recorded using RA/Dec coordinates must also include a date ("equinox") for those coordinates. Therefore the current star catalogues list their objects for the epoch and equinox J2000.0.

2. Polaris, the pole star, won't stay a good indicator of the location of the Northern celestial pole. In 14,000 years time Polaris will be nearly $47°$ away from the celestial pole!

3. The change in declination causes a shift in the rising and setting positions of the stars along the horizon. Figure 16.6 shows part of the horizon for latitude $\varphi = 30°$ North. For a given year (left vertical labels), make a horizontal line to find rising azimuth of the bright stars indicated by the twisting lines. Depending on where on the celestial sphere a star is located, it may appear to move north or south, or be almost stationary for several centuries.

16.5 Parallax

Parallax is the change of angular position of two stationary points relative to each other as seen by an observer, due to the motion of said observer. Or more simply put, it is the apparent shift of an object against a background due to a change in observer position.

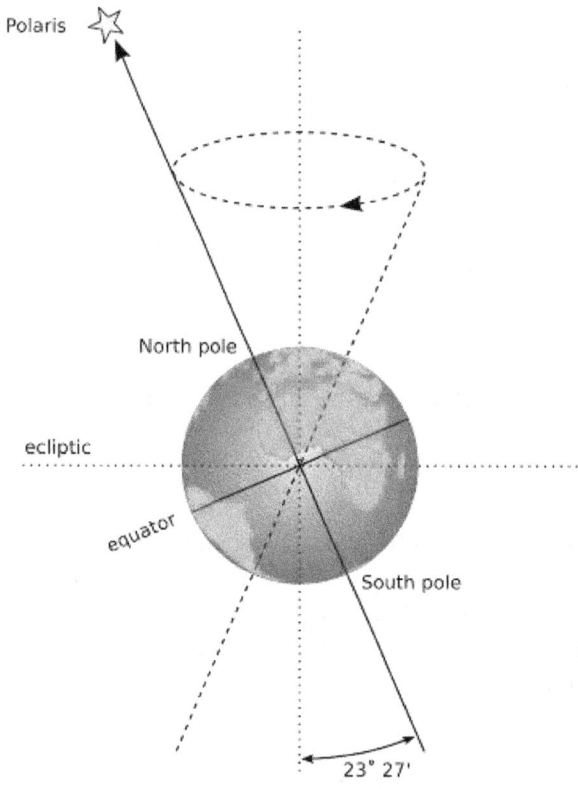

Figure 16.5: Precession

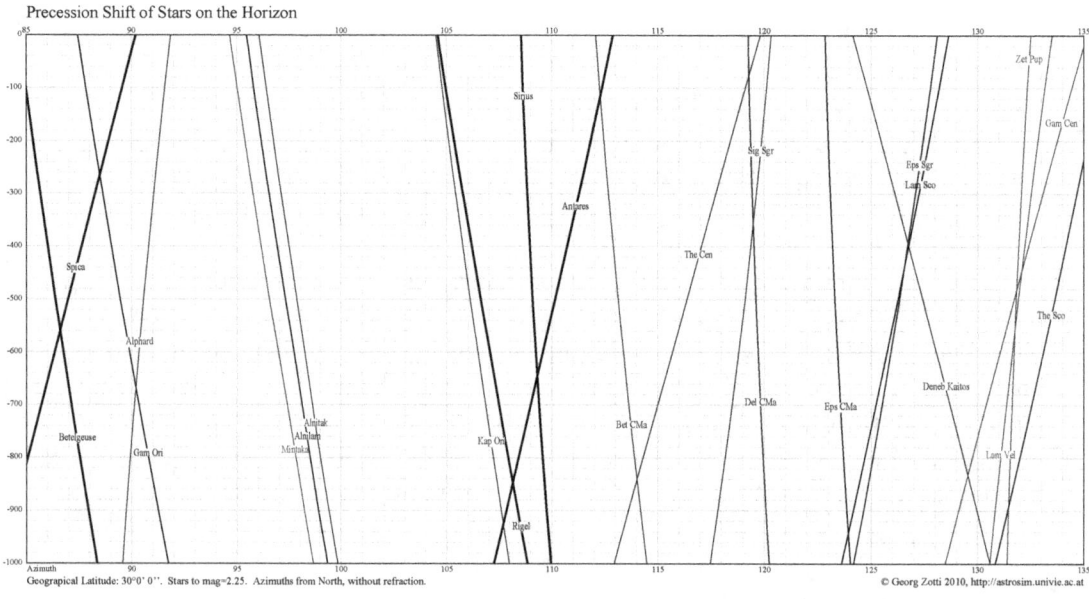

Figure 16.6: Precession: Change of rising positions of the stars along the eastern horizon from azimuths 85 to 135 degrees, between years 1000 BC and 0, for latitude $\varphi = 30°$.

This can be demonstrated by holding ones thumb up at arm's length. Closing one eye, note the position of the thumb against the background. After swapping which eye is open (without moving), the thumb appears to be in a different position against the background.

16.5.1 Geocentric and Topocentric Observations

When computing planetary positions was done manually by adding numbers tabulated in yearly almanacs, computing the Earth's position and, say, position of a minor planet was usually good enough to find the object in the sky. In both cases, the exact numbers refer to the gravitational centres of the respective bodies. However, we are sitting on Earth's surface, so the observed planet will be seen in a slightly shifted location. The amount for objects in the inner solar system is usually just a few arcseconds and is mostly negligible when we just want to find an object. But it makes a difference when it comes to observations of stellar occultations by planets or asteroids. Such a body may measure only a few tens of kilometres, and the shadow track which it leaves on Earth's surface is of approximately the same size.[10]

A much closer and bigger object is the Moon, which can also occult stars. It can even occult the one big star we call the Sun: this is a Solar Eclipse. And here it makes a huge difference where on the planet you are located.

If you are interested in astronomical computing, you may still be interested in geocentric numerical results. From the Settings panel (F2), tab "Tools", there is a checkbox for "Topocentric Coordinates". Switch it off to put yourself into the center of the planet you are located.

16.5.2 Stellar Parallax

A similar thing happens due to the Earth's motion around the Sun. Nearby stars appear to move against more distant background stars, as illustrated in figure 16.7. The movement of nearby stars against the background is called *stellar parallax*, or *annual parallax*.

Since we know the distance the radius of the Earth's orbit around the Sun from other methods, we can use simple geometry to calculate the distance of the nearby star if we measure annual parallax.

As can be seen from figure 16.7, the annual parallax p is half the angular distance between the apparent positions of the nearby star. The distance of the nearby object is d. Astronomers use a unit of distance called the parsec (pc) which is defined as the distance at which a nearby star has $p = 1''$.

Even the nearest stars exhibit very small movement due to parallax. The closest star to the Earth other than the Sun is Proxima Centauri. It has an annual parallax of $0.77199''$, corresponding to a distance of $1.295\,\mathrm{pc}$ (4.22 light years).

Even with the most sensitive instruments for measuring the positions of the stars it is only possible to use parallax to determine the distance of stars up to about 1,600 light years from the Earth, after which the annual parallax is so small it cannot be measured accurately enough.

In Stellarium, the annual parallax can be listed in the object information for stars when available. It is not used for the positional calculations.

16.6 Proper Motion

Proper motion is the change in the position of a star over time as a result of its motion through space relative to the Sun. It does not include the apparent shift in position of star due to annular parallax. The star exhibiting the greatest proper motion is *Barnard's Star* which moves more than ten seconds of arc per year.

---

[10]Unfortunately Stellarium (as of V0.15) is not accurate enough to reliably compute such occultations. Even a deviation of 0.5 arcseconds is too much here.

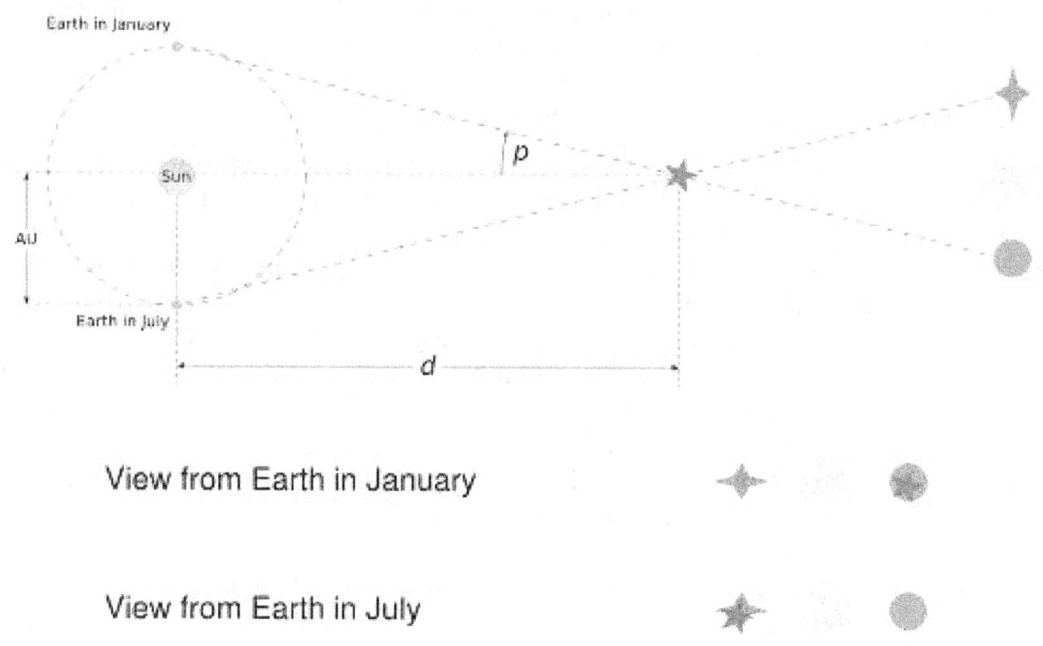

View from Earth in January

View from Earth in July

Figure 16.7: Stellar Parallax

If you want to simulate the effect of proper motion with Stellarium, put the map into equatorial view mode, switch off ground and cardinal marks, and set some high time lapse speed. You will see a few stars change their locations quite soon, those are usually stars in our galactic neighbourhood.

Note however some limitations:

1. Stellarium will stop at ±100.000 years. This limit may be still suitable for the stellar locations. The planetary locations are not trustworthy outside of a much closer temporal window (see section E). You cannot simulate the sky over the dinosaurs or such things.

2. Proper motion is only modelled by the linear components. True 3D motion in space requires more computation, which would slow down the program.

3. Double stars are listed in catalogs as two individual stars with their current proper motion. They may be seen flying apart, which is of course not realistic.

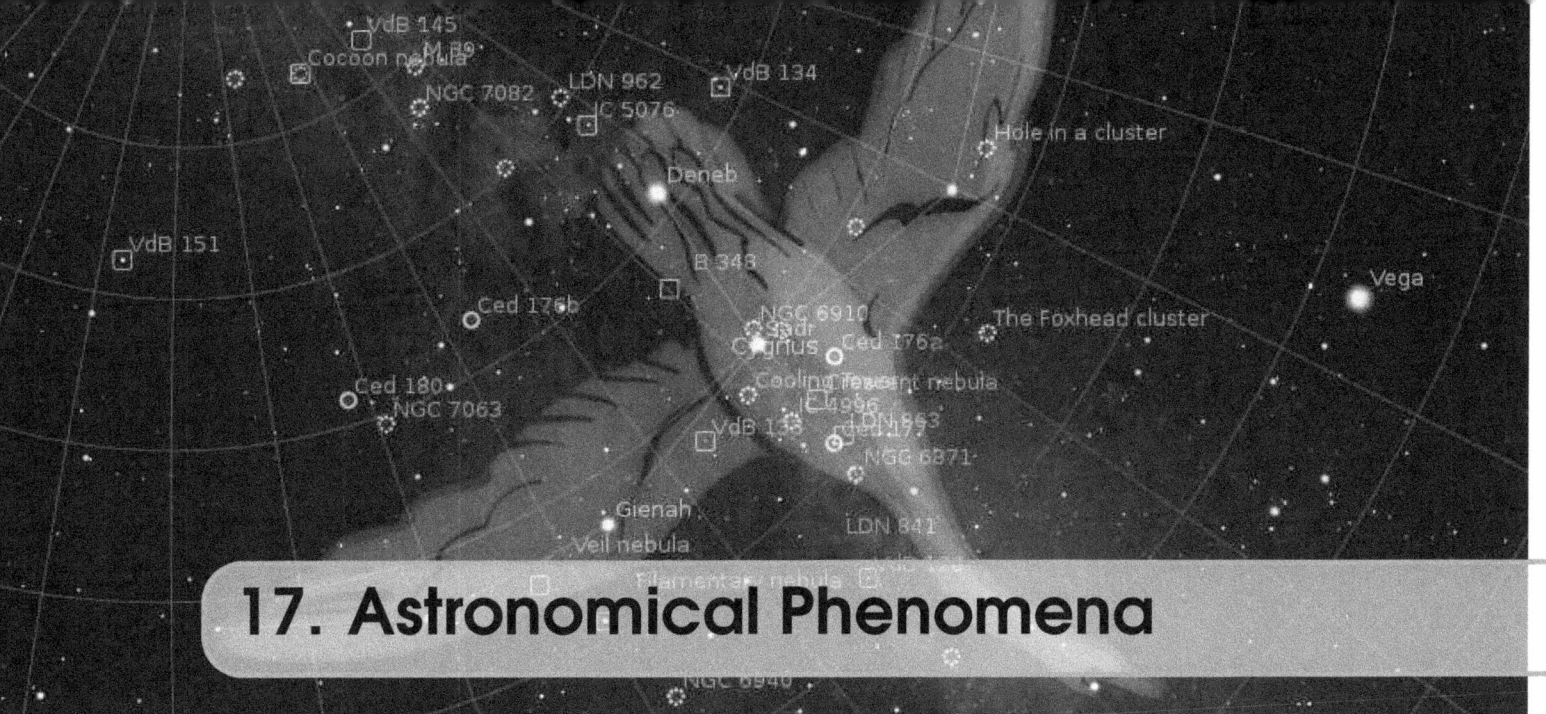

17. Astronomical Phenomena

BARRY GERDES, WITH ADDITIONS BY GEORG ZOTTI

This chapter focuses on the observational side of astronomy — what we see when we look at the sky.

17.1 The Sun

Without a doubt, the most prominent object in the sky is the Sun. The Sun is so bright that when it is in the sky, its light is scattered by the atmosphere to such an extent that almost all other objects in the sky are rendered invisible.

The Sun is a star like many others but it is much closer to the Earth at approximately 150 million kilometres (a distance also called 1 Astronomical Unit). The next nearest star, Proxima Centauri is approximately 260,000 times further away from us than the Sun! The Sun is also known by its Latin name, *Sol*.

Over the course of a year, the Sun appears to move round the celestial sphere in a great circle known as the *ecliptic*. Stellarium can draw the ecliptic on the sky. To toggle drawing of the ecliptic, press the ⌑,⌑ key.

WARNING: Looking at the Sun can permanently damage the eye. Never look at the Sun without using the proper filters! By far the safest way to observe the Sun it to look at it on a computer screen, courtesy of Stellarium!

17.2 Stars

The Sun is just one of billions of stars. Even though many stars have a much greater absolute magnitude than the Sun (they give out more light), they have an enormously smaller apparent magnitude due to their large distance. Stars have a variety of forms — different sizes, brightnesses, temperatures, and colours. Measuring the position, distance and attributes of the stars is known as *astrometry*, and is a major part of observational astronomy.

17.2.1 Multiple Star Systems

Many stars have stellar companions. As many as six stars can be found orbiting one-another in close associations known as *multiple star systems — binary systems* being the most common with two stars. Multiple star systems are more common than solitary stars, putting our Sun in the minority group.

Sometimes multiple stars orbit each other in a way that means one will periodically eclipse the other. These are *eclipsing binaries* or *Algol variables*.

Optical Doubles & Optical Multiples

Sometimes two or more stars appear to be very close to one another in the sky, but in fact have great separation, being aligned from the point of view of the observer but of different distances. Such pairings are known as *optical doubles* and *optical multiples*.

17.2.2 Constellations

The constellations are groupings of stars that are visually close to one another in the sky. The actual groupings are fairly arbitrary — different cultures have grouped stars together into different constellations. In many cultures, the various constellations have been associated with mythological entities. As such people have often projected pictures into the skies as can be seen in figure 17.1 which shows the constellation of Ursa Major. On the left is a picture with the image of the mythical Great Bear, on the right only a line-art version (or *stick figure*) is shown. The seven bright stars of Ursa Major are widely recognised, known variously as "the plough", the "pan-handle", and the "big dipper". This sub-grouping is known as an *asterism* — a distinct grouping of stars. On the right, the picture of the bear has been removed and only a constellation diagram remains.

Figure 17.1: Ursa Major

Stellarium can draw both constellation diagrams and artistic representations of the constellations. Multiple sky cultures are supported: Western, Polynesian, Egyptian, Chinese, and several other sky cultures are available, although at time of writing the non-Western constellations are not complete, and as yet there are no artistic representations of these sky-cultures.

Aside from historical and mythological value, to the modern astronomer the constellations provide a way to segment the sky for the purposes of describing locations of objects, indeed one of the first tasks for an amateur observer is *learning the constellations* — the process of becoming familiar with the relative positions of the constellations, at what time of year a constellation is visible, and in which constellations observationally interesting objects reside. Internationally, astronomers have adopted 88 "Western" constellations as a common system for segmenting the sky. They are based on Greek/Roman mythology, but with several additions from Renaissance and later centuries. As such some formalisation has been adopted, each constellation having a *proper name*, which is in Latin, and a three letter abbreviation of that name. For example, Ursa Major has the abbreviation UMa. Also, each "Western" constellation has clearly defined boundaries, which you can draw in Stellarium when you press the [B] key[1]. On the other hand, the shapes of mythological figures, and also stick figures, have not been canonized, so you will find deviations between Stellarium and printed atlases.

17.2.3 Star Names

Stars can have many names. The brighter stars often have *common names* relating to mythical characters from the various traditions. For example the brightest star in the sky, Sirius is also known as The Dog Star (the name Canis Major — the constellation Sirius is found in — is Latin for "The Great Dog").

Most bright names have been given names in antiquity. PTOLEMY's most influential book, the *Syntaxis*, was translated to the Arab language in the age of early Muslim scientists. When, centuries later, the translation, called *Almagest*, was re-introduced to the re-awakening European science, those names, which often only designated the position of the star within the figure, were taken from the books, often misspelled, and used henceforth as proper names.

A few more proper names have been added later, sometimes dedicatory names added by court astronomers into their maps. There are also 3 stars named after the victims of the Apollo 1 disaster in 1967. Today, the International Astronomical Union (IAU) is the only scientifically accepted authority which can give proper names to stars. Some companies offer a paid name service for commemoration or dedication of a star for deceased relatives or such, but all you get here is a piece of paper with coordinates of (usually) an unremarkably dim star only visible in a telescope, and a name to remember, stored (at best) in the company's database.

There are several more formal naming conventions that are in common use.

Bayer Designation

German astronomer JOHANN BAYER devised one such system for his atlas, the *Uranographia*, first published in 1603. His scheme names the stars according to the constellation in which they lie prefixed by a lower case Greek letter, starting at α for (usually) the brightest star in the constellation and proceeding with $\beta, \gamma, \ldots$ in descending order of apparent magnitude. For example, such a *Bayer Designation* for Sirius is "α Canis Majoris" (note that the genitive form of the constellation name is used; today also the short form α CMa is in use). There are some exceptions to the descending magnitude ordering, and some multiple stars (both real and optical) are named with a numerical superscript after the Greek letter, e.g. $\pi^1 \ldots \pi^6$ Orionis.

Flamsteed Designation

English astronomer JOHN FLAMSTEED numbered stars in each constellation in order of increasing right ascension followed by the genitive form of the constellation name, for example "61 Cygni" (or short: "61 Cyg").

Hipparcos

Hipparcos (for High Precision Parallax Collecting Satellite) was an astrometry mission of the European Space Agency (ESA) dedicated to the measurement of stellar parallax and the proper motions of stars. The project was named in honour of the Greek astronomer HIPPARCHUS.

---

[1]These boundaries or borders have been drawn using star maps from 1875. Due to the effect of *precession*, these borders are no longer parallel to today's coordinates.

Ideas for such a mission dated from 1967, with the mission accepted by ESA in 1980. The satellite was launched by an Ariane 4 on 8 August 1989. The original goal was to place the satellite in a geostationary orbit above the earth, however a booster rocket failure resulted in a highly elliptical orbit from 315 to 22,300 miles altitude. Despite this difficulty, all of the scientific goals were accomplished. Communications were terminated on 15 August 1993.

The program was divided in two parts: the *Hipparcos experiment* whose goal was to measure the five astrometric parameters of some 120,000 stars to a precision of some 2 to 4 milli arc-seconds and the *Tycho experiment*, whose goal was the measurement of the astrometric and two-colour photometric properties of some 400,000 additional stars to a somewhat lower precision.

The final Hipparcos Catalogue (120,000 stars with 1 milli arc-second level astrometry) and the final Tycho Catalogue (more than one million stars with 20-30 milli arc-second astrometry and two-colour photometry) were completed in August 1996. The catalogues were published by ESA in June 1997. The Hipparcos and Tycho data have been used to create the Millennium Star Atlas: an all-sky atlas of one million stars to visual magnitude 11, from the Hipparcos and Tycho Catalogues and 10,000 non-stellar objects included to complement the catalogue data.

There were questions over whether Hipparcos has a systematic error of about 1 milli arc-second in at least some parts of the sky. The value determined by Hipparcos for the distance to the Pleiades is about 10% less than the value obtained by some other methods. By early 2004, the controversy remained unresolved.

Stellarium uses the Hipparcos Catalogue for star data, as well as having traditional names for many of the brighter stars. The stars tab of the search window allows for searching based on a Hipparcos Catalogue number (as well as traditional names), e.g. the star Sadalmelik in the constellation of Aquarius can be found by searching for the name, or its Hipparcos number, 109074.

Figure 17.2 shows the information Stellarium displays when a star is selected. At the top, the common name, Bayer/Flamsteed designations and Hipparcos number are shown, followed by the RA/Dec coordinates, apparent magnitude, distance and other data.

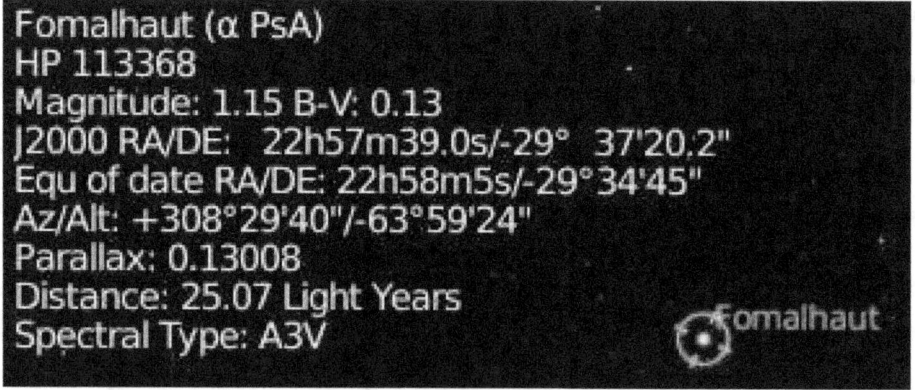

Figure 17.2: Star Names and Data

17.2.4 Spectral Type & Luminosity Class

Stars have many different colours. Seen with the naked eye most appear to be white, but this is due to the response of the eye — at low light levels the eye is not sensitive to colour. Typically the unaided eye can start to see differences in colour only for stars that have apparent magnitude brighter than 1. Betelgeuse, for example has a distinctly red tinge to it, and Sirius appears to be blue, while Vega is the prototype "white" star.

By splitting the light from a star using a prism attached to a telescope and measuring the relative intensities of the colours of light the star emits — the *spectrum* — a great deal of interesting information can be discovered about a star including its surface temperature, and the presence of various elements in its atmosphere.

Spectral Type	Surface Temperature (K)	Star Colour
O	28,000—50,000	Blue
B	10,000—28,000	Blue-white
A	7,500—10,000	White-blue
F	6,000—7,500	Yellow-white
G	4,900—6,000	Yellow
K	3,500—4,900	Orange
M	2,000—3,500	Red

Table 17.1: Spectral Types

Astronomers groups stars with similar spectra into *spectral types*, denoted by one of the following letters: O, B, A, F, G, K and M.[2] Type O stars have a high surface temperature (up to around 50,000 K) while the at other end of the scale, the M stars are red and have a much cooler surface temperature, typically 3000 K. The Sun is a type G star with a surface temperature of around 5,500 K. Spectral types may be further sub-divided using a numerical suffix ranging from 0-9 where 0 is the hottest and 9 is the coolest. Table 17.1 shows the details of the various spectral types.

For about 90% of stars, the absolute magnitude increases as the spectral type tends to the O (hot) end of the scale. Thus the whiter, hotter stars tend to have a greater luminosity. These stars are called *main sequence* stars. There are however a number of stars that have spectral type at the M end of the scale, and yet they have a high absolute magnitude. These stars are close to the ends of their lives and have a very large size, and consequently are known as *giants*, the largest of these known as *super-giants*.

There are also stars whose absolute magnitude is very low regardless of the spectral class. These are known as *dwarf stars*, among them *white dwarfs* (dying stars) and *brown dwarfs* ("failed stars").

The *luminosity class* is an indication of the type of star — whether it is main sequence, a giant or a dwarf. Luminosity classes are denoted by a number in roman numerals, as described in table 17.2.

Luminosity class	Description
Ia, Ib	Super-giants
II	Bright giants
III	Normal giants
IV	Sub-giants
V	Main sequence
VI	Sub-dwarfs
VII	White-dwarfs

Table 17.2: Luminosity Classification

Plotting the luminosity of stars against their spectral type/surface temperature gives a diagram

---

[2]The classic mnemonic for students of astrophysics says: "Oh, Be A Fine Girl, Kiss Me".

called a Hertzsprung-Russell diagram (after the two astronomers EJNAR HERTZSPRUNG and HENRY NORRIS RUSSELL who devised it). A slight variation of this is shown in figure 17.3 (which is technically a colour/magnitude plot).

17.2.5 Variable Stars

Most stars are of nearly constant luminosity. The Sun is a good example of one which goes through relatively little variation in brightness (usually about 0.1% over an 11 year solar cycle). Many stars, however, undergo significant variations in luminosity, and these are known as *variable stars*. There are many types of variable stars falling into two categories, *intrinsic* and *extrinsic*.

Intrinsic variables are stars which have intrinsic variations in brightness, that is, the star itself gets brighter and dimmer. There are several types of intrinsic variables, probably the best-known and most important of which is the Cepheid variable whose luminosity is related to the period with which its brightness varies. Since the luminosity (and therefore absolute magnitude) can be calculated, Cepheid variables may be used to determine the distance of the star when the annual parallax is too small to be a reliable guide. This is especially welcome because they are giant stars, and so they are even visible in neighboring galaxies.

Extrinsic variables are stars of constant brightness that show changes in brightness as seen from the Earth. These include rotating variables, stars whose apparent brightness change due to rotation, and eclipsing binaries.

17.3 Our Moon

The Moon is the large satellite which orbits the Earth approximately every 28 days. It is seen as a large bright disc in the early night sky that rises later each day and changes shape into a crescent until it disappears near the Sun. After this it rises during the day then gets larger until it again becomes a large bright disc again.

17.3.1 Phases of the Moon

As the moon moves round its orbit, the amount that is illuminated by the sun as seen from a vantage point on Earth changes. The result of this is that approximately once per orbit, the moon's face gradually changes from being totally in shadow to being fully illuminated and back to being in shadow again. This process is divided up into various phases as described in table 17.3.

New Moon	The moon's disc is fully in shadow, or there is just a slither of illuminated surface on the edge.
Waxing Crescent	Less than half the disc is illuminated, but more is illuminated each night.
First Quarter	Approximately half the disc is illuminated, and increasing each night.
Waxing Gibbous	More than half of the disc is illuminated, and still increasing each night.
Full Moon	The whole disc of the moon is illuminated.
Waning Gibbous	More than half of the disc is illuminated, but the amount gets smaller each night.
Last Quarter	Approximately half the disc is illuminated, but this gets less each night.
Waning Crescent	Less than half the disc of the moon is illuminated, and this gets less each night.

Table 17.3: Lunar Phases

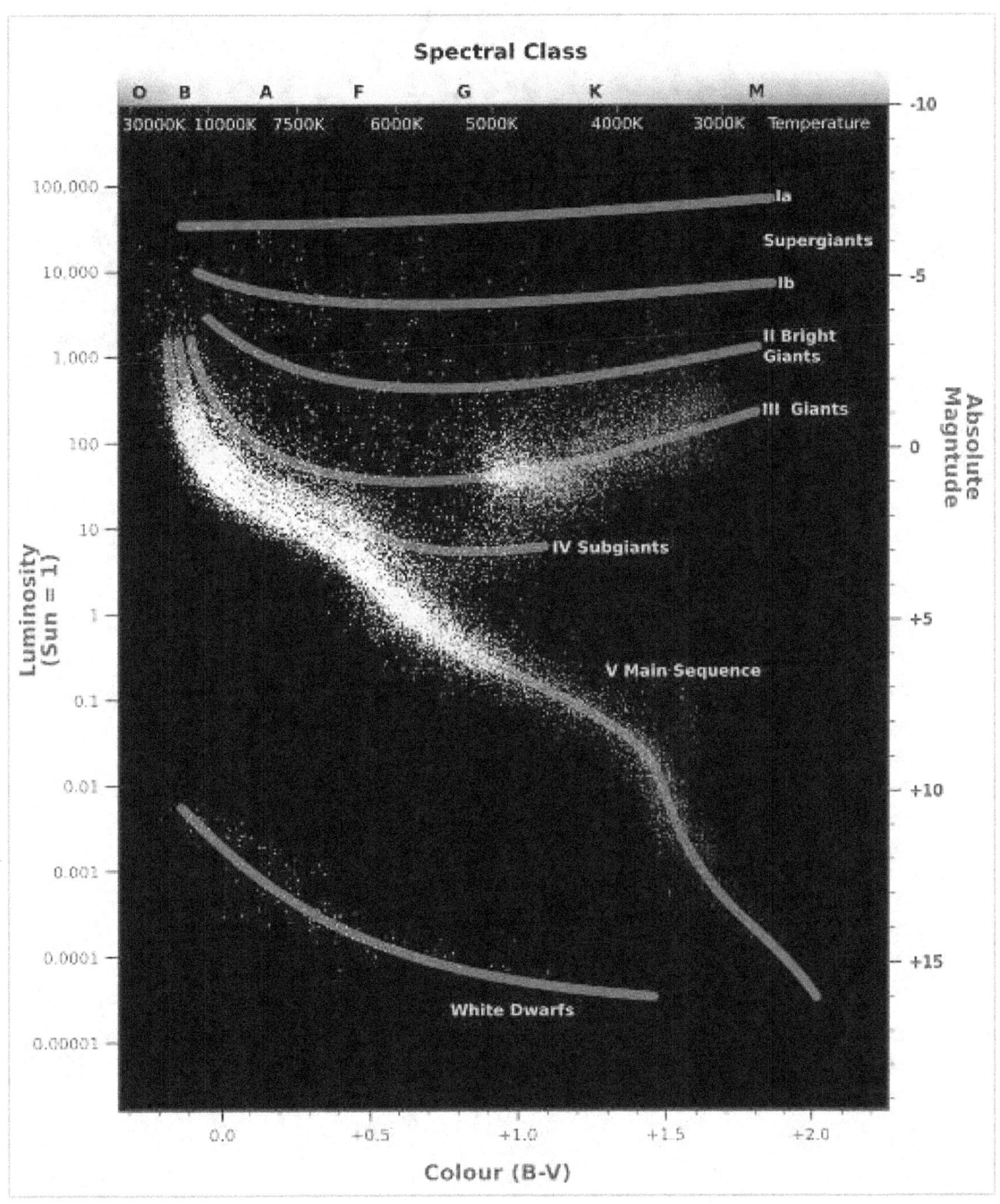

Figure 17.3: Hertzsprung-Russell Diagram

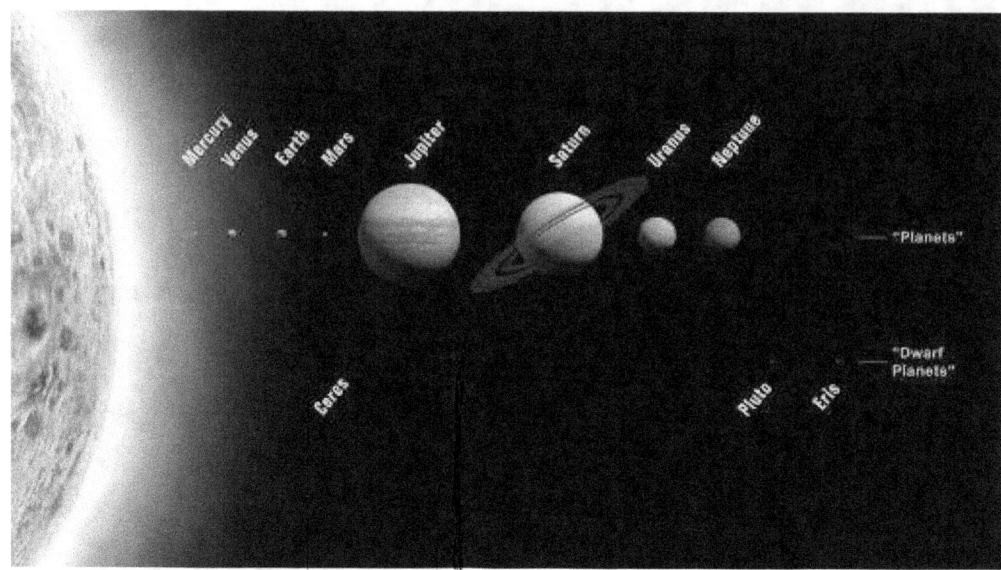

Figure 17.4: The Planets

17.4 The Major Planets

Unlike the stars whose relative positions remain more or less constant, the planets seem to move across the sky over time (the word "planet" comes from the Greek for "wanderer"). The planets are siblings of the Earth, massive bodies that are in orbit around the Sun. Until 2006 there was no formal definition of a planet, leading to some confusion about the classification for some bodies traditionally regarded as being planets, but which didn't seem to fit with the others.

In 2006 the International Astronomical Union defined a planet as a celestial body that, within the Solar System:

1. is in orbit around the Sun
2. has sufficient mass for its self-gravity to overcome rigid body forces so that it assumes a hydrostatic equilibrium (nearly round) shape; and
3. has cleared the neighbourhood around its orbit

or within another system:

1. is in orbit around a star or stellar remnants
2. has a mass below the limiting mass for thermonuclear fusion of deuterium; and
3. is above the minimum mass/size requirement for planetary status in the Solar System.

Moving from the Sun outwards, the 8 major planets are: Mercury, Venus, Earth, Mars, Jupiter, Saturn, Uranus and Neptune. Since the formal definition of a planet in 2006 Pluto has been relegated to having the status of *dwarf planet*, along with bodies such as Ceres and Eris. See figure 17.4.

17.4.1 Terrestrial Planets

The planets closest to the sun are called collectively the *terrestrial planets*. The terrestrial planets are: Mercury, Venus, Earth and Mars.

The terrestrial planets are relatively small, comparatively dense, and have solid rocky surface. Most of their mass is made from solid matter, which is mostly rocky and/or metallic in nature.

17.4.2 Jovian Planets

Jupiter, Saturn, Uranus and Neptune make up the *Jovian planets*, also called *gas giants*. They are much more massive than the terrestrial planets, and do not have a solid surface. Jupiter is the largest of all the planets with a diameter of about 12, and mass over 300 times that of the Earth!

The Jovian planets do not have a solid surface – the vast majority of their mass being in gaseous form (although they may have rocky or metallic cores). Because of this, they have an average density which is much less than the terrestrial planets. Saturn's mean density is only about $0.7\,\mathrm{g/cm^3}$ – it would float in water!

17.5 The Minor Bodies

As well as the Major Planets, the solar system also contains innumerable smaller bodies in orbit around the Sun. These are generally the *dwarf planets* (Ceres, Pluto, Eris), the other *minor planets*, also known as *planetoids* or *asteroids*, and comets.

17.5.1 Asteroids

Asteroids are celestial bodies orbiting the Sun in more or less regular orbits mostly between Mars and Jupiter. They are generally rocky bodies like the inner (terrestrial) planets, but of much smaller size. They are countless in number ranging in size from about ten meters to hundreds of kilometres.

17.5.2 Comets

A comet is a small body in the solar system that orbits the Sun and (at least occasionally) exhibits a coma (or atmosphere) and/or a tail.

Most comets have a very eccentric orbit (featuring a highly flattened ellipse, or even a parabolic track), and as such spend most of their time a very long way from the Sun. Comets are composed of rock, dust and ices. When they come close to the Sun, the heat evaporates the ices, causing a gaseous release. This gas and loose material which comes away from the body of the comet is swept away from the Sun by the Solar wind, forming the tail.

Most larger comets exhibit two kinds of tail: a straight gas tail (often blue-green in photographs), and a wider, occasionally curved dust tail (reflecting whitish sunlight).

Comets whose orbit brings them close to the Sun more frequently than every 200 years are considered to be *short period* comets, the most famous of which is probably Comet Halley, named after the British astronomer EDMUND HALLEY, which has an orbital period of roughly 76 years.

17.6 Meteoroids

These objects are small pieces of space debris left over from the early days of the solar system that orbit the Sun. They come in a variety of shapes, sizes an compositions, ranging from microscopic dust particles up to about ten meters across.

Sometimes these objects collide with the Earth. The closing speed of these collisions is generally extremely high (tens of kilometres per second). When such an object ploughs through the Earth's atmosphere, a large amount of kinetic energy is converted into heat and light, and a visible flash or streak can often be seen with the naked eye. Even the smallest particles can cause these events which are commonly known as *shooting stars*.

While smaller objects tend to burn up in the atmosphere, larger, denser objects can penetrate the atmosphere and strike the surface of the planet, sometimes leaving meteor craters.

Sometimes the angle of the collision means that larger objects pass through the atmosphere but do not strike the Earth. When this happens, spectacular fireballs are sometimes seen.

Meteoroids is the name given to such objects when they are floating in space.

A *Meteor* is the name given to the visible atmospheric phenomenon.

Meteorites is the name given to objects that penetrate the atmosphere and land on the surface.

In some nights over the year you can observe increased meteorite activity. Those meteors seem to come from a certain point in the sky, the *Radiant*. But what we see is similar to driving through a mosquito swarm which all seem to come head-on. Earth itself moves through space, and sweeps up a dense cloud of particles which originates from a comet's tail. Stellarium's Meteor Shower plugin (see section 12.6) can help you planning your next meteor observing night.

17.7 Zodiacal Light and *Gegenschein*

In very clear nights on the best observing sites, far away from the light pollution of our cities, you can observe a feeble glow also known as "false twilight" after evening twilight in the west, or before dawn in the east. The glow looks like a wedge of light along the ecliptic. Exactly opposite the sun, there is another dim glow that can be observed with dark-adapted eyes in perfect skies: the *Gegenschein* (counterglow).

This is sunlight reflected off the same dust and meteoroids in the plane of our solar system which is the source of meteors. Stellarium's sky can show the Zodiacal light [28], but observe how quickly light pollution kills its visibility!

17.8 The Milky Way

There is a band of very dense stars running right round the sky in huge irregular stripe. Most of these stars are very dim, but the overall effect is that on very dark clear nights we can see a large, beautiful area of diffuse light in the sky. It is this for which we name our galaxy the *Milky Way*.

The reason for this effect is that our galaxy is somewhat like a disc, and we are off to one side. Thus when we look towards the centre of the disc, we see more a great concentration of stars (there are more star in that direction). As we look out away from the centre of the disc we see fewer stars - we are staring out into the void between galaxies!

It's a little hard to work out what our galaxy would look like from far away, because when we look up at the night sky, we are seeing it from the inside. All the stars we can see are part of the Milky Way, and we can see them in every direction. However, there is some structure. There is a higher density of stars in particular places.

17.9 Nebulae

Seen with the naked eye, binoculars or a small telescope, a *nebula* (plural *nebulae*) is a fuzzy patch on the sky. Historically, the term referred to any extended object, but the modern definition excludes some types of object such as galaxies.

Observationally, nebulae are popular objects for amateur astronomers – they exhibit complex structure, spectacular colours (in most cases only visible in color photography) and a wide variety of forms. Many nebulae are bright enough to be seen using good binoculars or small to medium sized telescopes, and are a very photogenic subject for astro-photographers.

Nebulae are associated with a variety of phenomena, some being clouds of interstellar dust and gas in the process of collapsing under gravity, some being envelopes of gas thrown off during a supernova event (so called *supernova remnants*), yet others being the remnants of dumped outer layers around dying stars (*planetary nebulae*).

Examples of nebulae for which Stellarium has images include the Crab Nebula (M1), which is a supernova remnant, and the Dumbbell Nebula (M27) and the Ring Nebula (M57) which are planetary nebulae.

17.9.1 The Messier Objects

The *Messier* objects are a set of astronomical objects catalogued by CHARLES MESSIER in his catalogue of *Nebulae and Star Clusters* first published in 1774. The original motivation behind the catalogue was that Messier was a comet hunter, and was frustrated by objects which resembled but were not comets. He therefore compiled a list of these annoying objects.

The first edition covered 45 objects numbered M1 to M45. The total list consists of 110 objects, ranging from M1 to M110. The final catalogue was published in 1781 and printed in the *Connaissance des Temps* in 1784. Many of these objects are still known by their Messier number.

Because the Messier list was compiled by astronomers in the Northern Hemisphere, it contains only objects from the north celestial pole to a celestial latitude of about $-35°$. Many impressive Southern objects, such as the Large and Small Magellanic Clouds are excluded from the list. Because all of the Messier objects are visible with binoculars or small telescopes (under favourable conditions), they are popular viewing objects for amateur astronomers. In early spring, astronomers sometimes gather for "Messier Marathons", when all of the objects can be viewed over a single night.

Stellarium includes images of many Messier objects.

17.10 Galaxies

Stars, it seems, are gregarious – they like to live together in groups. These groups are called galaxies. The number of stars in a typical galaxy is literally astronomical – many *billions* – sometimes over *hundreds of billions* of stars!

Our own star, the sun, is part of a galaxy. When we look up at the night sky, all the stars we can see are in the same galaxy. We call our own galaxy the Milky Way (or sometimes simply "the Galaxy"[3]).

Other galaxies appear in the sky as dim fuzzy blobs. Only four are normally visible to the naked eye. The Andromeda galaxy (M31) visible in the Northern hemisphere, the two Magellanic clouds, visible in the Southern hemisphere, and the home galaxy Milky Way, visible in parts from north and south under dark skies.

There are thought to be billions of galaxies in the universe comprised of an unimaginably large number of stars.

The vast majority of galaxies are so far away that they are very dim, and cannot be seen without large telescopes, but there are dozens of galaxies which may be observed in medium to large sized amateur instruments. Stellarium includes images of many galaxies, including the Andromeda galaxy (M31), the Pinwheel Galaxy (M101), the Sombrero Galaxy (M104) and many others.

Astronomers classify galaxies according to their appearance. Some classifications include *spiral galaxies*, *elliptical galaxies*, *lenticular galaxies* and *irregular galaxies*.

17.11 Eclipses

Eclipses occur when an apparently large celestial body (planet, moon etc.) moves between the observer (that's you!) and a more distant object – the more distant object being eclipsed by the nearer one.

17.11.1 Solar Eclipses

Solar eclipses occur when our Moon moves between the Earth and the Sun. This happens when the inclined orbit of the Moon causes its path to cross our line of sight to the Sun. In essence it is the

---

[3]Which means closely the same thing, the word deriving from Greek *gala*=Milk.

observer falling under the shadow of the moon.

There are three types of solar eclipses:

Partial The Moon only covers part of the Sun's surface.

Total The Moon completely obscures the Sun's surface.

Annular The Moon is at aphelion (furthest from Earth in its elliptic orbit) and its disc is too small to completely cover the Sun. In this case most of the Sun's disc is obscured – all except a thin ring around the edge.

17.11.2 Lunar Eclipses

Lunar eclipses occur when the Earth moves between the Sun and the Moon, and the Moon is in the Earth's shadow. They occur under the same basic conditions as the solar eclipse but can occur more often because the Earth's shadow is so much larger than the Moon's.

Total lunar eclipses are more noticeable than partial eclipses because the Moon moves fully into the Earth's shadow and there is very noticeable darkening. However, the Earth's atmosphere refracts light (bends it) in such a way that some sunlight can still fall on the Moon's surface even during total eclipses. In this case there is often a marked reddening of the light as it passes through the atmosphere, and this can make the Moon appear a deep red colour.

17.12 Observing Hints

When stargazing, there's a few little things which make a lot of difference, and are worth taking into account.

Dark skies For many people getting away from light pollution isn't an easy thing. At best it means a drive away from the towns, and for many the only chance to see a sky without significant glow from street lighting is on vacation. If you can't get away from the cities easily, make the most of it when you are away.

Wrap up warm The best observing conditions are the same conditions that make for cold nights, even in the summer time. Observing is not a strenuous physical activity, so you will feel the cold a lot more than if you were walking around. Wear a lot of warm clothing, don't sit/lie on the floor (at least use a camping mat, consider taking a deck-chair), and take a flask of hot drink.

Dark adaptation The true majesty of the night sky only becomes apparent when the eye has had time to become accustomed to the dark. This process, known as dark adaptation, can take up to half an hour, and as soon as the observer sees a bright light they must start the process over. Red light doesn't compromise dark adaptation as much as white light, so use a red torch if possible (and one that is as dim as you can manage with). A dim single red LED light is ideal, also to have enough light to take notes.

The Moon Unless you're particularly interested in observing the Moon on a given night, it can be a nuisance—it can be so bright as to make observation of dimmer objects such as nebulae impossible. When planning what you want to observe, take the phase and position of the Moon into account. Of course Stellarium is the ideal tool for finding this out!

Averted vision A curious fact about the eye is that it is more sensitive to dim light towards the edge of the field of view. If an object is slightly too dim to see directly, looking slightly off to the side but concentrating on the object's location can often reveal it.

Angular distance Learn how to estimate angular distances. Learn the angular distances described in section 16.3.4. If you have a pair of binoculars, find out the angular distance across the field of view and use this as a standard measure.

17.13 Atmospheric effects

17.13.1 Atmospheric Refraction

Atmospheric Refraction is a lifting effect of our atmosphere which can be observed by the fact that objects close to the horizon appear higher than they should be if computed only with spherical trigonometry. Stellarium simulates refraction for terrestrial locations when the atmosphere is switched on. Refraction depends on air pressure and temperature. Figure 17.5 has been created from the same formulae that are employed in Stellarium. You can see how fast refraction grows very close to the mathematical horizon.

Note that these models can only give approximate conditions. There are many weird effects in the real atmosphere, when temperature inversion layers can create light ducts, cause double sunsets etc.

Also note that the models give meaningful results only for altitudes above approximately $-2°$. Below that, in nature, there is always ground which blocks our view. In Stellarium you can switch off the ground, and you can observe a sunset with a strange egg-shaped sun below the horizon. This is of course nonsense. Stellarium is also not able to properly recreate the atmospheric distortions as seen from a stratosphere balloon, where the height of earth's surface is several degrees below the mathematical horizon.

17.13.2 Atmospheric Extinction

Atmospheric Extinction is the attenuation of light of a celestial body by Earth's atmosphere. In the last split-second of its travel into our eyes or detectors, light from outer space has to pass our atmosphere, through layers of mixed gas, water vapour and dust. If a star is in the zenith, its light must pass one *air mass* and is reduced by whatever amount of water and dust is above you. When the star is on the horizon, it has to pass about 40 times longer through the atmosphere: 40 air masses (Fig. 17.6. The number of air masses increases fast in low altitudes, this is why we see so few stars along the horizon. Usually blue light is extinguished more, this is why the sun and moon (and brigher stars) appear reddish on the horizon.

Stellarium can simulate extinction, and you can set the opacity of your atmosphere with a global factor k, the *magnitude loss per airmass* (see section 4.4.1). The best mountaintop sites may have $k = 0.15$, while $k = 0.25$ seems a value usable for good locations in lower altitudes.

17.13.3 Light Pollution

An ugly side effect of civilisation is a steady increase in outdoor illumination. Many people think it increases safety, but while this statement can be questioned, one definite result, aside from environmental issues like dangers for the nocturnal fauna, are ever worsening conditions for astronomical observations or just enjoyment of the night sky.

Stellarium can simulate light pollution, which is controlled from the light pollution section of the *Sky* tab of the *View* window. Light pollution levels are set using a numerical value between 1 and 9 which corresponds to the *Bortle Dark Sky Scale* (see Appendix B). In addition, local variations of the amount of light pollution can be included in a light pollution layer in the landscapes, see section 7 for details.

raction

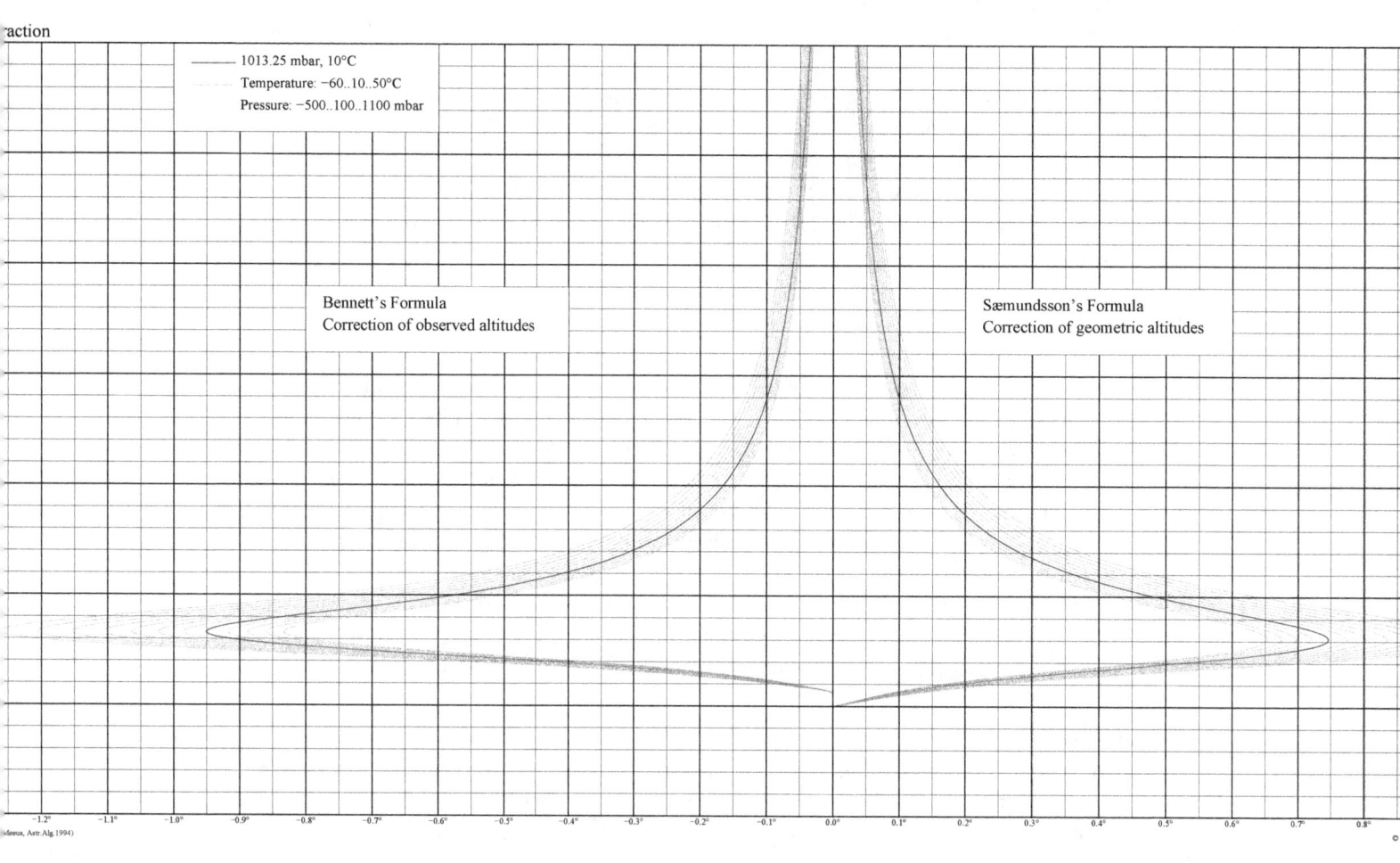

1013.25 mbar, 10°C
Temperature: −60..10..50°C
Pressure: −500..100..1100 mbar

Bennett's Formula
Correction of observed altitudes

Sæmundsson's Formula
Correction of geometric altitudes

−1.2° −1.1° −1.0° −0.9° −0.8° −0.7° −0.6° −0.5° −0.4° −0.3° −0.2° −0.1° 0.0° 0.1° 0.2° 0.3° 0.4° 0.5° 0.6° 0.7° 0.8°

Meeus, Astr.Alg.1994)

re 17.5: Refraction. The figure shows corrective values (degrees) which are subtracted from observed altitudes (left side) to reach
ides, or values to be added to computed values (right side). The models used are not directly inverse operations.

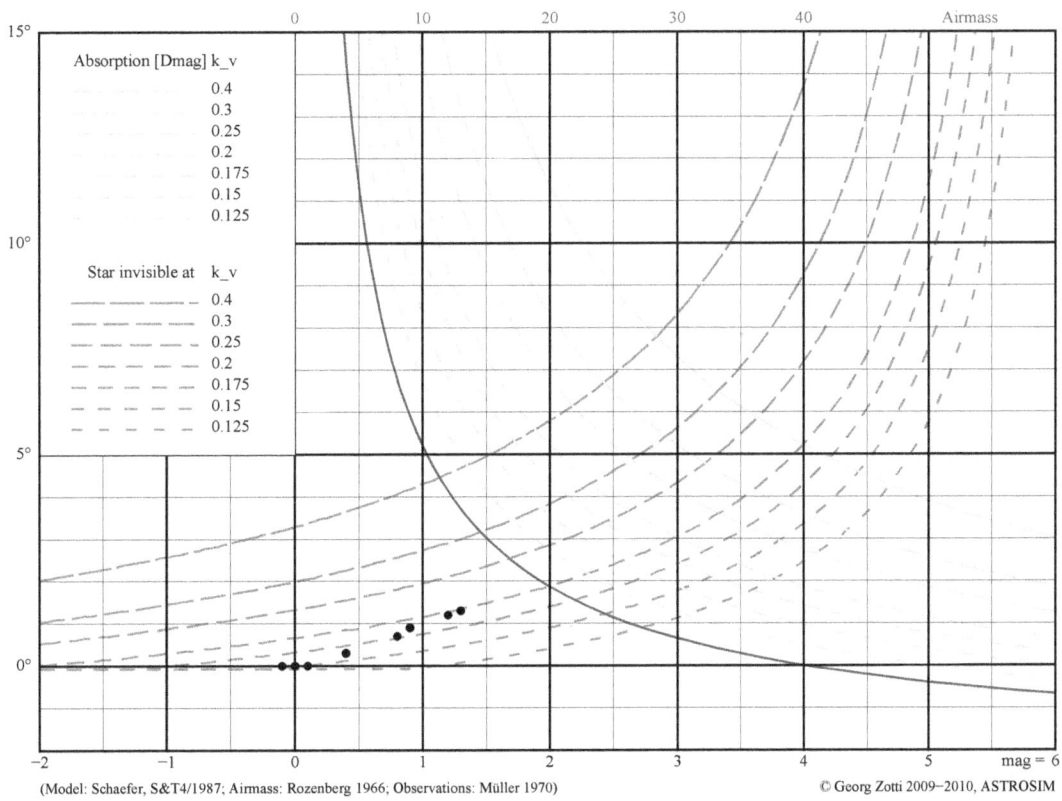

Figure 17.6: Airmass and Extinction. The figure shows Airmass (blue) along the line of sight in the altitude labeled on the left side. The green curves show how many magnitudes an object is dimmed down, depending on extinction factor k (called k_v in the figure). The red curves indicate at which altitude a star of given magnitude can be seen with good eyesight, again depending on k. The black dots are observed values found in the literature.

18. A Little Sky Guide

PAUL ROBINSON, WITH ADDITIONS BY ALEXANDER WOLF

This chapter lists some astronomical objects that can be located using Stellarium. All of them can be seen with the naked eye or binoculars. Since many astronomical objects have more than one name (often having a "proper name", a "common name" and various catalogue numbers), the chapter lists the name as it appears in Stellarium — use this name when using Stellarium's search function — and any other commonly used names.

The Location Guide entry gives brief instructions for finding each object using nearby bright stars or groups of stars when looking at the real sky — a little time spent learning the major constellations visible from your latitude will pay dividends when it comes to locating fainter (and more interesting!) objects. When trying to locate these objects in the night sky, keep in mind that Stellarium displays many stars that are too faint to be visible without optical aid, and even bright stars can be dimmed by poor atmospheric conditions and light pollution.

18.1 Dubhe and Merak, The Pointers

Type: Stars
Magnitude: 1.83, 2.36
Location Guide: *The two "rightmost" of the seven stars that form the main shape of "The Plough" (or "Big Dipper", part of Ursa Major).*

Northern hemisphere observers are very fortunate to have two stars that point towards Polaris which lie very close to the northern celestial pole. Whatever the time of night or season of the year they are always an immediate clue to the location of Polaris, the Pole Star.

18.2 M31, Messier 31, The Andromeda Galaxy

Type: Spiral Galaxy
Magnitude: 3.4

Location Guide: *Find the three bright stars that constitute the main part of the constellation of Andromeda. From the middle of these look toward the constellation of Cassiopeia.*

M31 is the most distant object visible to the naked eye, and among the few nebulae that can be seen without a telescope or powerful binoculars. Under good conditions it appears as a large fuzzy patch of light. It is a galaxy containing billions of stars whose distance is roughly 2.5 million light years from Earth.

18.3   The Garnet Star, μ Cephei

Type: Variable Star
Magnitude: 4.25 (Avg.)
Location Guide: *Cephius lies "above" the W-shape of Cassiopeia. The Garnet Star lies slightly to one side of a point halfway between 5 Cephei and 21 Cephei.*

A "supergiant" (1035 Solar radii) of spectral class M with a strong red colour. Given its name by Sir William Herschel in the 18th century, the colour is striking in comparison to its blue-white neighbours.

18.4   4 and 5 Lyrae, ε Lyrae

Type: Double Star
Magnitude: 4.7
Location Guide: *Close to Vega (α Lyrae), one of the brightest stars in the sky.*

In binoculars ε Lyrae is resolved into two separate stars. Remarkably, each of these is also a double star (although this will only be seen with a telescope) and all four stars form a physical system.

18.5   M13, Hercules Cluster

Type: Globular Cluster
Magnitude: 5.8
Location Guide: *Located approximately 1/3 of the way along a line from 44 (η) to 40 (ζ) Herculis.*

This cluster of hundreds of thousands of mature stars appears as a circular "cloud" using the naked eye or binoculars (a large telescope is required to resolve individual stars). Oddly the cluster appears to contain one young star and several areas that are almost devoid of stars.

18.6   M45, The Pleiades, The Seven Sisters

Type: Open Cluster
Magnitude: 1.2 (Avg.)
Location Guide: *Lies on the Bull's back, about 1/3 between Aldebaran in Taurus and Almaak in Andromeda.*

Depending upon conditions, six to 9 of the blueish stars in this famous cluster will be visible to someone with average eyesight, and in binoculars it is a glorious sight. The cluster has more than 500 members in total, many of which are shown to be surrounded by nebulous material in long exposure photographs.

18.7 Algol, The Demon Star, β Persei

Type: Variable Star
Magnitude: 3.0 (Avg.)
Location Guide: *Halfway between Aldebaran in Taurus and the middle star of the "W" of Cassiopeia.*

Once every three days or so, Algol's brightness changes from 2.1 to 3.4 and back within a matter of hours. The reason for this change is that Algol has a dimmer giant companion star, with an orbital period of about 2.8 days, that causes a regular partial eclipse. Although Algol's fluctuations in magnitude have been known since at least the 17th century, it was the first to be proved to be due to an eclipsing companion — it is therefore the prototype Eclipsing Variable.

18.8 Sirius, α Canis Majoris

Type: Star
Magnitude: -1.47
Location Guide: *Sirius is easily found by following the line of three stars in Orion's belt southwards.*

Sirius is a white dwarf star at a comparatively close 8.6 light years. This proximity and its high innate luminance makes it the brightest star in our sky. Sirius is a double star; its companion is a White Dwarf, much dimmer but very hot, and is believed to be smaller than the earth.

18.9 M44, The Beehive, Praesepe

Type: Open Cluster
Magnitude: 3.7
Location Guide: *Cancer lies about halfway between the twins (Castor & Pollux) in Gemini and Regulus, the brightest star in Leo. The Beehive can be found between Asellus Borealis and Asellus Australis.*

There are probably 350 or so stars in this cluster, although it appears to the naked eye simply as a misty patch. It contains a mixture of stars from red giants to white dwarf and is estimated to be some 700 million years old.

18.10 27 Cephei, δ Cephei

Type: Variable Star
Magnitude: 4.0 (Avg.)
Location Guide: *Locate the four stars that form the square of Cepheus. One corner of the square has two other bright stars nearby forming a distinctive triangle — δ is at the head of this triangle in the direction of Cassiopeia.*

δ Cephei gives its name to a whole class of variables, all of which are pulsating high-mass stars in the later stages of their evolution. δ Cephei is also a double star with a companion of magnitude 6.3 visible in binoculars.

18.11 M42, The Great Orion Nebula

Type: Nebula
Magnitude: 4.0
Location Guide: *Almost in the middle of the area bounded by Orion's belt and lower stars, Saiph and Rigel.*

The Great Orion Nebula is the brightest nebula visible in the night sky and lies at about 1.500 light years from earth. It is a truly gigantic gas and dust cloud that extends for several hundred light years, reaching almost halfway across the constellation of Orion. The nebula contains a cluster of hot young stars known as the Trapezium, and more stars are believed to be forming within the cloud.

18.12   La Superba, Y Canum Venaticorum, HIP 62223

Type: Star
Magnitude: 5.4 (Avg.)
Location Guide: *Almost the center of the arch of stars of Ursa Major's tail. Forms a neat triangle with Phekda (γ) and Alkaid (η, tail tip) in Ursa Major towards Canes Venatici.*

La Superba (215 Solar radii) is a "Carbon Star" — a group of relatively cool gigantic (usually variable) stars that have an outer shell containing high levels of carbon. This shell is very efficient at absorbing short wavelength blue light, giving carbon stars a distinctive red or orange tint. One of the coolest and reddest known stars.

18.13   52 and 53 Bootis, v^1 and v^2 Bootis

Type: Double Star
Magnitude: 5.0, 5.0
Location Guide: *Follow a line from Seginus (γ Boo, left shoulder) to Nekkar (β Boo, the head) and then continue for the same distance again to arrive at this double star.*

This optical double star consists of a pair of different spectral type, and 52 Bootis, at approximately 800 light years, is twice as far away as 53.

18.14   PZ Cas, HIP 117078

Type: Variable Star
Magnitude: 8.2 (Avg.)
Location Guide: *Lies about 1/3 between Caph (β Cas, the top right star of "W") in Cassiopeia and ι Cephei (32 Cep, top left star in "rectangle" of Cepheus).*

This faint red star is one of the biggest known stars — its average size parameter is 1565 Solar radii (the true value is from 1340 to 1940 Solar radii). PZ Cas is a pulsating variable star located in a region with heavy dust extinction.

18.15   VV Cephei, HIP 108317

Type: Variable Star, Double Star
Magnitude: 5.1 (Avg.)
Location Guide: *Lies near the center of the "rectangle" of Cepheus).*

This is an interesting eclipsing binary system — VV Cep A (1457 Solar radii) is a highly distorted star in a binary system, losing mass to its B-type companion VV Cephei B (10 Solar radii) for at least part of its orbit.

18.16   AH Scorpii, HIP 84071

Type: Variable Star
Magnitude: 7.1 (Avg.)

Location Guide: *Lies in Scorpius, about 1¼ of a continuation of the Scorpion's "tail".*

AH Sco (1411 Solar radii) is variable by nearly 3 magnitudes in the visual range, and an estimated 20% in total luminosity. The variation in diameter is not clear because the temperature also varies.

18.17 Albireo, β Cygni

Type: Double Star
Magnitude: 3.4, 5.1
Location Guide: *The "head" of Cygni.*

When viewed with the naked eye, it appears to be a single star. However, in a telescope it readily resolves into a double star, consisting of Albireo A (amber), and Albireo B (blue-green). Separated by 35″, the two components provide one of the best contrasting double stars in the sky due to their different colors.

Figure 18.1: Albireo: A Bright and Beautiful Double. *Credit & Copyright: Richard Yandrick.*

18.18 31 and 32 Cygni, o¹ and o² Cygni

Type: Multiple Star
Magnitude: 3.8, 4.0
Location Guide: *The two bright stars about 1/2 between Deneb (α Cygni, "tail" of Cygnus) and Rukh (δ Cygni, middle of the left "wing" of Cygnus).*

The wide binary o¹ Cygni and o² Cygni is separated 61′ (NNE), and this star is an easy naked-eye double. o¹ Cygni with 30 Cyg and HD 192579 (HIP 99676) is a moderately difficult triple system. And surprise — each of o¹ Cyg and o² Cyg are also Algol-type eclipsing binaries!

18.19 The Coathanger, Brocchi's Cluster, Cr 399

Type: Asterism
Location Guide: *It is best found by slowly sweeping across the Milky Way along an imaginary line from the bright star Altair toward the even brighter star Vega (around 1/3 between Altair and Vega).*

The asterism is made up of 10 stars ranging from 5th to 7th magnitude which form the conspicuous "coathanger", a straight line of 6 stars with a "hook" of 4 stars on the south side. Under a dark sky, Collinder 399 can be seen with the naked eye as an unresolved patch of light; binoculars or a telescope at very low power are usually needed in order to view the "coathanger" asterism.

18.20   Kemble's Cascade

Type: Asterism
Location Guide: *The asterism lies in the constellation Camelopardalis, about 1/3 between CS Cam and α Cam — HIP 18505 has magnitude 5 and can be found in the center of the chain.*
   Kemble's Cascade is a chain of stars which are visible in binocular even in light-polluted skies.

18.21   The Double Cluster, χ and h Persei, NGC 884 and NGC 869

Type: Open Clusters
Location Guide: *The two open clusters near stars χ and h in the constellation Perseus.*
   The Double Cluster (also known as Caldwell 14 or C14) is the common name for the naked-eye open clusters NGC 869 and NGC 884 (often designated h Persei and χ Persei, respectively). The Double Cluster is approximately the radiant of the Perseid meteor shower, which peaks annually around August 12 or 13. Although easy to locate in the northern sky, observing the Double Cluster in its two parts requires optical aid. They are described as being an "awe-inspiring" and "breathtaking" sight, and are often cited as targets in astronomical observers' guides.

18.22   Large Magellanic Cloud, PGC 17223

Type: Dwarf Spiral Galaxy
Magnitude: 0.9
Location Guide: *It is visible as a faint "cloud" in the night sky of the southern hemisphere straddling the border between the constellations of Dorado and Mensa, and it appears from Earth more than 20 times the width of the full moon.*
   The *Large Magellanic Cloud* (LMC) is a satellite galaxy of the Milky Way. The Large Magellanic Cloud is usually considered an irregular galaxy. However, it shows signs of a bar structure, and has more recently been reclassified as a Magellanic-type dwarf spiral galaxy. At a distance of 50 kiloparsecs, the LMC is the third closest galaxy to the Milky Way, after the Sagittarius Dwarf Spheroidal (~16 kiloparsecs) and the putative Canis Major Dwarf Galaxy (~12.9 kiloparsecs, though its status as a galaxy is under dispute) lying closer to the center of the Milky Way. The LMC has a diameter of about 14,000 light-years (~4.3 kpc) and a mass of approximately 10 billion (10^{10}) solar masses, making it roughly 1/100 as massive as the Milky Way. The LMC has a wide range of galactic objects and phenomena — surveys of the galaxy have found roughly 60 globular clusters, 400 planetary nebulae, and 700 open clusters, along with hundreds of thousands of giant and supergiant stars.

18.23   Tarantula Nebula, C 103, NGC 2070

Type: Nebula
Magnitude: 7.3
Location Guide: *Part of Large Magellanic Cloud.*
   The Tarantula Nebula (also known as 30 Doradus) is an H II region in the Large Magellanic Cloud (LMC). 30 Doradus has at its centre the star cluster NGC 2070 which includes the compact

Figure 18.2: Small and Large Magellanic Clouds over Paranal Observatory. *Credit: ESO/J. Colosimo.*

concentration of stars known as R136 that produces most of the energy that makes the nebula visible. The estimated mass of the cluster is 450,000 solar masses, suggesting it will likely become a globular cluster in the future [7]. The closest supernova observed since the invention of the telescope, Supernova 1987A, occurred in the outskirts of the Tarantula Nebula.

18.24 Small Magellanic Cloud, NGC 292, PGC 3085

Type: Dwarf Irregular Galaxy
Magnitude: 2.2
Location Guide: *It is located mostly in the constellation of Tucana and appears as a hazy, light patch in the night sky about 3° across, looking like a detached piece of the Milky Way. Since it has a very low surface brightness, it is best viewed from a dark site away from city lights.*

The *Small Magellanic Cloud* (SMC) is a dwarf galaxy near the Milky Way. It is classified as a dwarf irregular galaxy. It has a diameter of about 7,000 light-years and has a total mass of approximately 7 billion times the mass of the Sun. The SMC contains a central bar structure and it is speculated that it was once a barred spiral galaxy that was disrupted by the Milky Way to become somewhat irregular. At a distance of about 200,000 light-years, it is one of the Milky Way's nearest neighbors. It is also one of the most distant objects that can be seen with the naked eye.

18.25 ω Centauri cluster, C 80, NGC 5139

Type: Globular Cluster
Magnitude: 5.3
Location Guide: *It is located around 1/3 between ζ and γ Centauri.*

Omega Centauri (ω Cen), or NGC 5139, is a globular cluster in the constellation of Centaurus that was first identified as a non-stellar object by Edmond Halley in 1677. Located at a distance of 15,800 light-years (4,850 pc), it is the largest globular cluster in the Milky Way at a diameter of roughly 150 light-years. It is estimated to contain approximately 10 million stars and a total mass equivalent to 4 million solar masses. Omega Centauri is so distinctive from the other galactic globular clusters that it is thought to have an alternate origin as the core remnant of a disrupted dwarf galaxy [47].

18.26   47 Tucanae, C 106, NGC 104

Type: Globular Cluster
Magnitude: 4.1
Location Guide: *It can be seen with the naked eye near Small Magellanic Cloud.*

It is the one of brightest globular cluster in the sky, and is noted for having a very bright and dense core. It is also one of the most massive globular clusters in the Milky Way Galaxy, containing millions of stars. The cluster appears roughly the size of the full moon in the sky under ideal conditions.

18.27   The Coalsack Nebula, C 99

Type: Dark Nebula
Location Guide: *Dark region between α and β Crux.*

The Coalsack Nebula is the most prominent dark nebula in the skies, easily visible to the naked eye as a dark patch silhouetted against the southern Milky Way. It is located at a distance of approximately 600 light years away from Earth. The Coalsack is important in Australian Aboriginal astronomy, and forms the head of the Emu in the sky in several Aboriginal cultures.

18.28   Mira, o Ceti, 68 Cet

Type: Variable Star
Magnitude: 2.0–10.1
Location Guide: *It is located around 1/3 between δ and θ Ceti.*

Mira (o Ceti) is a star estimated 200–400 light years away in the constellation Cetus. Mira is a binary star system that consists of a red giant (Mira A) undergoing mass loss and a high temperature white dwarf companion (Mira B) that is accreting mass from the primary. Such an arrangement of stars is known as a symbiotic system, and this is the closest such symbiotic pair to the Sun. Mira is the brightest periodic variable in the sky that is not visible to the naked eye for part of its cycle. Its distance is uncertain. Mira A is a well-known example of a category of long-period variable stars known as Mira variables, which are named after it.

18.29 α Persei Cluster, Cr 39, Mel 20

Type: Open Cluster
Magnitude: 1.2
Location Guide: *Around the white-yellow supergiant Mirfak, also known as α Persei in direction to δ Persei.*

The Alpha Persei Cluster, also known as Melotte 20 or Collinder 39, is an open cluster in the constellation of Perseus. To the naked eye, the cluster consists of several blue (spectral class B) stars.

Figure 18.3: The Coalsack Nebula taken by the Wide Field Imager on the MPG/ESO 2.2-metre telescope. *Credit: ESO*.

18.30 M7, The Ptolemy Cluster

Type: Open Cluster
Magnitude: 3.3
Location Guide: *The cluster is easily detectable with the naked eye, close to the "sting" of Scorpius — near the center of the line between Shaula (λ Scorpii) and Kaus Media (δ Sagittarii).*

M7 has been known since antiquity — it was first recorded by the 2nd-century Greek-Roman astronomer Ptolemy, who described it as a nebula in 130 AD. Telescopic observations of the cluster reveal about 80 stars within a field of view of 1.3° across. At the cluster's estimated distance of 980 light years this corresponds to an actual diameter of 25 light years.

18.31 M24, The Sagittarius Star Cloud

Type: Star Cloud
Magnitude: 4.6
Location Guide: *In the Milky Way region near Polis (μ Sagittarii).*

The Sagittarius Star Cloud is a star cloud in the constellation of Sagittarius, approximately 600 light years wide. The stars, clusters and other objects comprising M24 are part of the Sagittarius or Sagittarius-Carina arms of the Milky Way galaxy. Messier described M24 as a "large nebulosity containing many stars" and gave its dimensions as being some 1.5° across.

18.32   **IC 4665, The Summer Beehive Cluster**

Type: Open Cluster
Magnitude: 4.2
Location Guide: *Visible to the naked eye near Cebalrai (β Ophiuchi).*

IC 4665 is an open cluster in the constellation Ophiuchus which is easily visible in the smallest of telescopes and also with binoculars. From a sufficiently dark place it is also visible to the naked eye.

18.33   **The E Nebula, Barnard 142 and 143**

Type: Dark Nebula
Location Guide: *A well-defined dark area on a background of Milky Way near Tarazed (γ Aquilae).*

The "E" or "Barnard's E" Nebula (officially designated as B142 and B143) is a pair of dark nebulae in the constellation Aquila. It is a well-defined dark area on a background of the Milky Way. The size of the nebula is roughly 0.5°, and its distance from earth is estimated at about 2000 light years.

19. Exercises

PAUL ROBINSON, WITH ADDITIONS BY ALEXANDER WOLF

19.1 Find M31 in Binoculars

M31 – the *Andromeda Galaxy* – is the most distant object visible to the naked eye. Finding it in binoculars is a rewarding experience for new-comers to observing.

19.1.1 Simulation

1. Set the location to a mid-Northern latitude if necessary (M31 isn't always visible for Southern hemisphere observers). The UK is ideal.
2. Find M31 and set the time so that the sky is dark enough to see it. The best time of year for this at Northern latitudes is Autumn/Winter, although there should be a chance to see it at some time of night throughout the year.
3. Set the field of view to 6° (or the field of view of your binoculars if they're different. 6° is typical for 7x50 binoculars).
4. Practise finding M31 from the bright stars in Cassiopeia and the constellation of Andromeda. Learn the chain of stars that extends from Andromeda's central star perpendicular to her body.

19.1.2 For Real

This part is not going to be possible for many people. First, you need a good night and a dark sky. In urban areas with a lot of light pollution it's going to be very hard to see Andromeda.

19.2 Handy Angles

As described in section 16.3.4, your hand at arm's length provides a few useful estimates for angular size. It's useful to know whether your handy angles are typical, and if not, what they are.

The method here below is just one way to do it – feel free to use another method of your own construction!

Hold your hand at arm's length with your hand open – the tips of your thumb and little finger as far apart as you can comfortably hold them. Get a friend to measure the distance between your thumb and your eye, we'll call this D. There is a tendency to over-stretch the arm when someone is measuring it – try to keep the thumb-eye distance as it would be if you were looking at some distant object.

Without changing the shape of your hand, measure the distance between the tips of your thumb and little finger. It's probably easiest to mark their positions on a piece of paper and measure the distance between the marks, we'll call this d. Using some simple trigonometry, we can estimate the angular distance θ using equation (16.4).

Repeat the process for the distance across a closed fist, three fingers and the tip of the little finger.

For example, for one author $D = 72$ cm, $d = 21$ cm, so:

$$\theta = 2 \cdot \arctan\left(\frac{21}{144}\right) \approx 16\frac{1}{2}^{\circ} \qquad (19.1)$$

Remember that handy angles are not very precise – depending on your posture at a given time the values may vary by a fair bit.

19.3 Find a Lunar Eclipse

Stellarium comes with two scripts for finding lunar eclipses, but can you find one on a different date?

19.4 Find a Solar Eclipse

Find a Solar Eclipse using Stellarium and take a screenshot of it. Use the location panel and see how the eclipse look on different locations at the same time.

19.5 Find a retrograde motion of Mars

Find a Mars using Stellarium and take a series of screenshots of him position in different time. Use the date and time panel and see how a motion of Mars is changes in the time. Find a periods of direct and retrograde motion of Mars.

19.6 Analemma

Set a time as the noon and set time rate in pause, turn on the azimuthal grid. Find the Sun and check him horizontal coordinates. Use the date and time panel and see how the horizontal coordinates of the Sun is changes in time (please use one time step for simulation — look the position of the Sun every 7 days for example). Use the location panel and see how the positions of the Sun look on different location at the same times. Check change the positions of the Sun on Mars in same times.

19.7 Transit of Venus

Set date at 6th June 2012, find Venus near the Sun and change scale of the view. Find time of all four contacts and maximum of transit for your location. Because the Sun appears to rotate as it crossed at the sky, Venus will appear to move on some curve — for example it will be an inverted "U" shape for eastern states of Australia. Check difference of shape of path of Venus for equatorial and azimuthal mounts. Find few dates and times for transit of Venus in past and in future.

19.8 Transit of Mercury

Set date at 9th May 2016, find Mercury near the Sun and change scale of the view. Find time of all four contacts and maximum of transit for your location. Because the Sun appears to rotate as it crossed at the sky, Mercury will appear to move on some curve — for example it will be an inverted "U" shape for observers from Europe. Check difference of shape of path of Mercury for equatorial and azimuthal mounts. Find few dates and times for transit of Mercury in past and in future.

19.9 Triple shadows on Jupiter

Set date at 24th January 2015, find Jupiter and change scale of the view. Find the time when three shadows will be visible on disk of Jupiter. Which moons given those shadows? Check which moons give of triple shadows on Jupiter at 3 June 2014. Details for this phenomenon you can find in book "Mathematical astronomy morsels" by J. Meeus [36].

19.10 Jupiter without satellites

Set date at 9th November 2019, find Jupiter and change scale of the view. Find time limits for the disappearance of the moons. Which moons are on the back of planet, and which — in front of her? Please give answer on same questions for 3 September 2009. Details for this phenomenon you can find in book "Mathematical astronomy morsels" by J. Meeus [36]

19.11 Mutual occultations of planets

Set date at 13th October 1590, find Venus and change the scale of the view. Find time of minimal separation of Venus and Mars on this date. Set typical scale of view for visual observation and check Venus and Mars again. What would said the typical observer of that time? Find the minimal separation between Venus and Saturn near end of August 1771. What would said the typical observer on Earth? Find the minimal separation between Mars and Mercury in first half of August 2079.

Find mutual occultation of Earth and Mercury as seen from Mars near the end of November 2022. Find mutual occultation of Saturn and Mercury as seen from Venus near the middle of May 1967.

Details for this phenomenon you can find in books "More mathematical astronomy morsels" and "Mathematical astronomy morsels III" by J. Meeus [38, 39]

19.12 The proper motion of stars

Turn off ground and atmosphere, set time rate on pause, find the star HIP 87937 and change the scale of the view to better see of the Barnard's star. Use the date and time panel and see how the location of the star is changes in time (please try use one time step for simulation — look the position of the star every year for example). When Barnard's star will be cross the border of constellation? When Barnard's star has crossed the border of constellation in the past?

Please check the proper motions of Sirius (HIP 32349), Procyon (HIP 37279), 61 Cyg (HIP 104214) and τ Cet (HIP 8102). Which star has fastest proper motion? Which star has the slowest rate of proper motion?

When observer will be see mutual occultation of Arcturus and HIP 68706? Find the minimal angular separation between θ and ε Ind in the past.

How to will change the appearance of follow constellations on wide range of time (-100000..100000 years for example): Ursa Major, Orion, Bootes?

V

Appendices

A   Default Hotkeys 217

B   The Bortle Scale of Light Pollution 223

C   Star Catalogues 227

D   Configuration Files 249

E   Accuracy 275

F   Contributors 277

G   GNU Free Documentation License 279

     Bibliography 285

     Index 291

A. Default Hotkeys

The currently configured hotkeys are visible in the Help dialog (F1). Here is the default list for reference.

A.1 Display Options

Shortcut key	Description
P	Planets
Alt + P	Planet labels
Ctrl + P	Planet markers
O	Planet orbits
Shift + T	Planet trails
S	Stars
Alt + S	Stars labels
D	Deep-sky objects (symbols)
I	Deep-sky objects background images
M	Milky Way
Ctrl + Shift + Z	Zodiacal Light
G	Ground
F	Fog
Shift + G	Illumination
Ctrl + Shift + G	Landscape Labels
A	Atmosphere
Q	Cardinal points
,	Ecliptic line
.	Equator line
E	Equatorial grid

Shortcut key	Description
Z	Azimuthal grid
H	Horizon line
;	Meridian line
C	Constellation lines
V	Constellation labels
R	Constellation art
B	Constellation boundaries
Ctrl + Shift + N	Native planet names (from starlore)
F11	Full-screen mode
Ctrl + N	Night mode
Ctrl + Shift + H	Flip scene horizontally
Ctrl + Shift + V	Flip scene vertically

A.2   Miscellaneous

Shortcut key	Description
Ctrl + C	Copy selected object information to clipboard
Ctrl + Q	Quit
Ctrl + S	Save screenshot
Ctrl + M	Switch between equatorial and azimuthal mount
Ctrl + T	Toggle visibility of GUI

A.3   Movement and Selection

Shortcut key	Description
	Center on selected object
Ctrl + H	Go to home
Ctrl + G	Set home planet to selected planet ("go there")
Shift + E	Look towards East
Shift + N	Look towards North
Shift + S	Look towards South
Shift + W	Look towards West
Shift + Z	Look towards Zenith
T	Track object
/	Zoom in on selected object
\	Zoom out

A.4   Date and Time

Shortcut key	Description
=	Add 1 solar day
-	Subtract 1 solar day
Ctrl + =	Add 1 solar hour
Ctrl + -	Subtract 1 solar hour

Shortcut key	Description
Alt + =	Add 1 sidereal day
Alt + -	Subtract 1 sidereal day
)	Add 7 solar days
(	Subtract 7 solar days
Ctrl + Alt + Shift +)	Add 1 sidereal year
Ctrl + Alt + Shift + (	Subtract 1 sidereal year
J	Decrease time speed
Shift + J	Decrease time speed (a little)
K	Set normal time rate
L	Increase time speed
Shift + L	Increase time speed (a little)
7	Set time rate to zero
8	Set time to now
Ctrl + D , P	Pause script execution
Ctrl + D , R	Resume script execution
Ctrl + D , S	Stop script execution

A.5 Scripts

Shortcut key	Description
Ctrl + D , 0	Run script "Solar System Screensaver"
Ctrl + D , 1	Run script "Constellations Tour"
Ctrl + D , 2	Run script "Sky Culture Tour"
Ctrl + D , 3	Run script "Screensaver"

A.6 Windows

Shortcut key	Description
F1	Toggle Help window
F2	Toggle Configuration window
F3	Toggle Search window
F4	Toggle Sky and viewing options window
F5	Toggle Date/time window
F6	Toggle Location window
F7	Toggle Shortcuts window
F10	Toggle AstroCalc window
F12	Toggle Script Console window
Alt + B	Toggle Bookmarks window

A.7 Plugins

A.7.1 Angle Measure

Shortcut key	Description

Ctrl + A	Toggle angle measurement

A.7.2 ArchaeoLines

Shortcut key	Description
Ctrl + U	Toggle archaeolines

A.7.3 Compass Marks

Shortcut key	Description
Ctrl + Alt + C	Toggle compass marks

A.7.4 Equation of Time

Shortcut key	Description
Ctrl + Alt + T	Show solution for Equation of Time

A.7.5 Exoplanets

Shortcut key	Description
Ctrl + Alt + E	Show exoplanets
Alt + E	Toggle Exoplanets configuration window

A.7.6 Field of View

Shortcut key	Description
Ctrl + Alt + 1	Set FOV to 180°
Ctrl + Alt + 2	Set FOV to 90°
Ctrl + Alt + 3	Set FOV to 60°
Ctrl + Alt + 4	Set FOV to 45°
Ctrl + Alt + 5	Set FOV to 20°
Ctrl + Alt + 6	Set FOV to 10°
Ctrl + Alt + 7	Set FOV to 5°
Ctrl + Alt + 8	Set FOV to 2°
Ctrl + Alt + 9	Set FOV to 1°
Ctrl + Alt + 0	Set FOV to 0.5°

A.7.7 Meteor Showers

Shortcut key	Description
Ctrl + Shift + M	Toggle meteor showers
Shift + M	Toggle radiant labels
Ctrl + Alt + M	Show search dialog

Ctrl + Alt + Shift + M   Show settings dialog

A.7.8 Oculars

Shortcut key	Description
Ctrl + O	Ocular view
Alt + O	Oculars popup menu
Alt + C	Show crosshairs
Ctrl + B	Telrad sight

A.7.9 Pulsars

Shortcut key	Description
Ctrl + Alt + P	Show pulsars

A.7.10 Quasars

Shortcut key	Description
Ctrl + Alt + Q	Show quasars

A.7.11 Satellites

Shortcut key	Description
Ctrl + Z	Toggle satellite hints
Alt + Shift + Z	Toggle satellite labels
Alt + Z	Toggle Satellites configuration window

A.7.12 Scenery3d: 3D landscapes

Shortcut key	Description
Ctrl + W	Toggle 3D landscape
Ctrl + Shift + W	Show settings dialog
Ctrl + Alt + W	Show viewpoint dialog
Ctrl + R , T	Toggle location text
Ctrl + R , S	Toggle shadows
Ctrl + R , L	Toggle torchlight
Ctrl + R , D	Toggle debug information
Ctrl + R , P	Reload shaders

A.7.13 Solar System Editor

Shortcut key	Description
Ctrl + Alt + S	Import orbital elements in MPC format...

A.7.14 **Telescope Control**

Shortcut key	Description
Ctrl + 0	Move a telescope to a given set of coordinates
Ctrl + 1	Move telescope #1 to selected object
Alt + 1	Move telescope #1 to the point currently in the center of the screen
Alt + 2	Move telescope #2 to the point currently in the center of the screen
Alt + 3	Move telescope #3 to the point currently in the center of the screen
Alt + 4	Move telescope #4 to the point currently in the center of the screen
Alt + 5	Move telescope #5 to the point currently in the center of the screen
Alt + 6	Move telescope #6 to the point currently in the center of the screen
Alt + 7	Move telescope #7 to the point currently in the center of the screen
Alt + 8	Move telescope #8 to the point currently in the center of the screen
Alt + 9	Move telescope #9 to the point currently in the center of the screen
Ctrl + 2	Move telescope #2 to selected object
Ctrl + 3	Move telescope #3 to selected object
Ctrl + 4	Move telescope #4 to selected object
Ctrl + 5	Move telescope #5 to selected object
Ctrl + 6	Move telescope #6 to selected object
Ctrl + 7	Move telescope #7 to selected object
Ctrl + 8	Move telescope #8 to selected object
Ctrl + 9	Move telescope #9 to selected object

B. The Bortle Scale of Light Pollution

In Sky&Telescope February 2001 JOHN E. BORTLE published a sky quality scale which describes the amount of light pollution. Stellarium's light pollution setting tries to follow this scale. For completeness we reproduce from Wikipedia:

B.1 Excellent dark sky site

Level: 1
Limiting magnitude (eye): 7.6 – 8.0

Zodiacal light, gegenschein, zodiacal band visible; M33 direct vision naked-eye object; Scorpius and Sagittarius regions of the Milky Way cast obvious shadows on the ground; Airglow is readily visible; Jupiter and Venus affect dark adaptation; surroundings basically invisible.

B.2 Typical truly dark site

Level: 2
Limiting magnitude (eye): 7.1 – 7.5

Airglow weakly visible near horizon; M33 easily seen with naked eye; highly structured Summer Milky Way; distinctly yellowish zodiacal light bright enough to cast shadows at dusk and dawn; clouds only visible as dark holes; surroundings still only barely visible silhouetted against the sky; many Messier globular clusters still distinct naked-eye objects.

B.3 Rural sky

Level: 3
Limiting magnitude (eye): 6.6 – 7.0

Some light pollution evident at the horizon; clouds illuminated near horizon, dark overhead; Milky Way still appears complex; M15, M4, M5, M22 distinct naked-eye objects; M33 easily visible with averted vision; zodiacal light striking in spring and autumn, color still visible; nearer surroundings vaguely visible.

B.4 Rural/suburban transition

Level: 4
Limiting magnitude (eye): 6.1 – 6.5

Light pollution domes visible in various directions over the horizon; zodiacal light is still visible, but not even halfway extending to the zenith at dusk or dawn; Milky Way above the horizon still impressive, but lacks most of the finer details; M33 a difficult averted vision object, only visible when higher than 55°; clouds illuminated in the directions of the light sources, but still dark overhead; surroundings clearly visible, even at a distance.

B.5 Suburban sky

Level: 5
Limiting magnitude (eye): 5.6 – 6.0

Only hints of zodiacal light are seen on the best nights in autumn and spring; Milky Way is very weak or invisible near the horizon and looks washed out overhead; light sources visible in most, if not all, directions; clouds are noticeably brighter than the sky.

B.6 Bright suburban sky

Level: 6
Limiting magnitude (eye): 5.1 – 5.5

Zodiacal light is invisible; Milky Way only visible near the zenith; sky within 35° from the horizon glows grayish white; clouds anywhere in the sky appear fairly bright; surroundings easily visible; M33 is impossible to see without at least binoculars, M31 is modestly apparent to the unaided eye.

B.7 Suburban/urban transition

Level: 7
Limiting magnitude (eye): 5.0 at best

Entire sky has a grayish-white hue; strong light sources evident in all directions; Milky Way invisible; M31 and M44 may be glimpsed with the naked eye, but are very indistinct; clouds are brightly lit; even in moderate-sized telescopes the brightest Messier objects are only ghosts of their true selves.

B.8 City sky

Level: 8
Limiting magnitude (eye): 4.5 at best

Sky glows white or orange — you can easily read; M31 and M44 are barely glimpsed by an experienced observer on good nights; even with telescope, only bright Messier objects can be detected; stars forming familiar constellation patterns may be weak or completely invisible.

B.9   Inner City sky

Level: 9

Limiting magnitude (eye): 4.0 at best

Sky is brilliantly lit with many stars forming constellations invisible and many weaker constellations invisible; aside from Pleiades, no Messier object is visible to the naked eye; only objects to provide fairly pleasant views are the Moon, the Planets and a few of the brightest star clusters.

C. Star Catalogues

This chapter provides technical descriptions on how Stellarium records its star catalogues, and the related file formats.

C.1 Stellarium's Sky Model

C.1.1 Zones

The celestial sphere is split into zones, which correspond to the triangular faces of a geodesic sphere. The number of zones (faces) depends on the level of sub-division of this sphere. The lowest level, 0, is an icosahedron (20 faces), subsequent levels L of sub-division give the number of zones n as:

$$n = 20 \cdot 4^L \tag{C.1}$$

Stellarium uses levels 0 to 7 in the existing star catalogues. Star Data Records contain the position of a star as an offset from the central position of the zone in which that star is located, thus it is necessary to determine the vector from the observer to the centre of a zone, and add the star's offsets to find the absolute position of the star on the celestial sphere.

This position for a star is expressed as a 3-dimensional vector which points from the observer (at the centre of the geodesic sphere) to the position of the star as observed on the celestial sphere.

C.2 Star Catalogue File Format

C.2.1 General Description

Stellarium's star catalogue data is kept in the `stars/default` sub-directory of the Installation Directory and/or User Directory (see section 5.1).

The main catalogue data is split into several files:

`stars_0_0v0_6.cat`
`stars_1_0v0_6.cat`

```
stars_2_0v0_6.cat
stars_3_1v0_3.cat
stars_4_1v0_1.cat
stars_5_2v0_1.cat
stars_6_2v0_1.cat
stars_7_2v0_1.cat
stars_8_2v0_1.cat
```

There also exist some control and reference files:

```
stars_hip_cids_0v0_0.cat
stars_hip_sp_0v0_2.cat
gcvs_hip_part.dat
wds_hip_part.dat
cross-id.dat
stars.ini
name.fab
```

When Stellarium starts, it reads the `stars.ini` file, from which it determines the names of the other files, which it then loads.

The files `stars_hip_cids_0v0_0.cat` and `stars_hip_sp_0v0_2.cat` contain reference data for the main catalogue files. The file `gcvs_hip_part.dat` contains data about variables stars (see section C.3) and file `wds_hip_part.dat` contains data about double stars (see section C.4) which again contains references into the main catalogue files.

The file `cross-id.dat` (see section C.5) contains cross-identification data between HIP, SAO and HD designations.

A given catalogue file models stars for one and only one level (i.e. for a fixed number of zones), which is recorded in the header of the file. Individual star records do not contain full positional coordinates, instead they contain coordinates relative to the central position of the zone they occupy. Thus, when parsing star catalogues, it is necessary to know about the zone model to be able to extract positional data.

File	Data Type	Data Record Size	Geodesic Level	#Records	Notes
stars_0_0v0_3.cat	0	28 bytes	0	4,979	Hipparcos
stars_1_0v0_3.cat	0	28 bytes	1	21,800	Hipparcos
stars_2_0v0_3.cat	0	28 bytes	2	150,705	Hipparcos
stars_3_1v0_2.cat	1	10 bytes	3	428,466	Tycho 2
stars_4_1v0_0.cat	1	10 bytes	4	1,702,042	Tycho 2
stars_5_2v0_0.cat	2	8 bytes	5	7,083,058	NOMAD
stars_6_2v0_0.cat	2	8 bytes	6	24,670,782	NOMAD
stars_7_2v0_0.cat	2	8 bytes	7	50,733,321	NOMAD
stars_8_2v0_0.cat	2	8 bytes	7	92,304,337	NOMAD

For a given catalogue file, there may be one of three formats for the actual star data. The variation comes from the source of the data - the larger catalogues of fainter stars providing less data per star than the brighter star catalogues. See tables *Stellarium's star catalogue* and for details.

Stellarium's star catalogues based on Hipparcos [3, 17], Tycho 2 [24] and NOMAD [69] catalogues.

C.2.2 File Sections

The catalogue files are split into three main sections as described in table *File sections*.

Section	Offset	Description
File Header Record	0	Contains magic number, geodesic subdivision level, and magnitude range
Zone Records	32	A list of how many records there are for each zone. The length of the zones section depends on the level value from the header
Star Data Records	32+4n	This section of the file contains fixed-size star records, as described below. Records do not contain zone information, which must be inferred by counting how many records have been read so far and switching zones when enough have been read to fill the number of stars for the zone, as specified in the zones section above. The value of n used in the offset description is the number of zones, as described above.

C.2.3 Record Types

File Header Record

The File Header Record describes file-wide settings. It also contains a magic number which servers as a file type identifier. See table *Header Record*.

Name	Offset	Type	Size	Description
Magic	0	int	4	The magic number which identifies the file as a star catalogue. 0xde0955a3
Data Type	4	int	4	This describes the type of the file, which defines the size and structure of the Star Data record for the file.
Major Version	8	int	4	The file format major version number
Minor Version	12	int	4	The file format minor version number
Level	16	int	4	Sets the level of sub-division of the geodesic sphere used to create the zones. 0 means an icosahedron (20 triangular faces), subsequent levels of sub-division lead to numbers of zones as described in section Zones
Magnitude Minimum	20	int	4	The low bound of the magnitude scale for values in this file. Note that this is still an integer in Stellarium's own internal representation
Magnitude Range	24	int	4	The range of magnitudes expressed in this file
Magnitude Steps	28	int	4	The number of steps used to describes values in the range

Zone Records

The *Zone Records* section of the file lists the number of star records there are per zone. The number of zones is determined from the level value in the File Header Record, as described in section Zones. The Zones section is simply a list of integer values which describe the number of stars for each zone. The total length of the Zones section depends on the number of zones. See table *Zones section*.

Name	Offset	Type	Size	Description
num stars in zone 0	0	int	4	The number of records in this file which are in zone 0
num stars in zone 1	4	int	4	The number of records is this file which are in zone 1
...				
num stars in zone n	$4n$	int	4	The number of records is this file which are in zone n

Star Data Records

After the Zones section, the actual star data starts. The star data records themselves do not contain the zone in which the star belongs. Instead, the zone is inferred from the position of the record in the file. For example, if the Zone Records section of the file says that the first 100 records are for zone 0, the next 80 for zone 1 and so on, it is possible to infer the zone for a given record by counting how many records have been read so far.

The actual record structure depends on the value of the Data Type, as found in the File Header Record.

See tables *Star Data Record Type 0*, *Star Data Record Type 1* and *Star Data Record Type 2* for record structure details.

It should be noted that although the positional data loses accuracy as one progresses though the Star Record Types, this is compensated for by the face that the number of zones is much higher for the files where the smaller precision position fields are used, so the actual resolution on the sky isn't significantly worse for the type 1 and 2 records in practice.

Name	Offset	Type	Size	Description
hip	0	int	3	Hipparcos catalogue number
component_ids	3	unsigned char	1	This is an index to an array of catalogue number suffixes. The list is read from the **stars_hip_component_ids.cat** file. The value of this field turns out to be the line number in the file - 1
x0	4	int	4	This is the position of the star relative to the central point in the star's zone, in axis 1
x1	8	int	4	This is the position of the star relative to the central point in the star's zone, in axis 2
b_v	9	unsigned char	1	This is the magnitude level in B-V colour. This value refers to one of 256 discrete steps in the magnitude range for the file
mag	10	unsigned char	1	This is the magnitude level in the V-I colour. This value refers to one of 256 discrete steps in the magnitude range for the file

sp_int	11	unsigned short int	2	This is the index in an array of spectral type descriptions which is taken from the file stars_hip_sp.cat, the index corresponds to the line number in the file - 1
dx0	13	int	4	This is the proper motion of the star in axis 1
dx1	17	int	4	This is the proper motion of the star in axis 2
plx	21	int	4	This is the parallax of the star. To get the actual value, divide by 10000.

Name	Offset	Type	Size	Description
x0	0	int	20 bits	This is the position of the star relative to the central point in the star's zone, in axis 1
x1	20 bits	int	20 bits	This is the position of the star relative to the central point in the star's zone, in axis 2
dx0	40 bits	int	14 bits	This is the proper motion of the star in axis 1
dx1	54 bits	int	14 bits	This is the proper motion of the star in axis 2
b_v	68 bits	unsigned int	7 bits	This is the magnitude level in B-V colour. This value refers to one of 256 discrete steps in the magnitude range for the file
mag	75 bits	unsigned int	5 bits	This is the magnitude level in the V-I colour. This value refers to one of 256 discrete steps in the magnitude range for the file

Name	Offset	Type	Size	Description
x0	0	int	18 bits	This is the position of the star relative to the central point in the star's zone, in axis 1
x1	18 bits	int	18 bits	This is the position of the star relative to the central point in the star's zone, in axis 2
b_v	36 bits	unsigned int	7 bits	This is the magnitude level in B-V colour. This value refers to one of 256 discrete steps in the magnitude range for the file
mag	43 bits	unsigned int	5 bits	This is the magnitude level in the V-I colour. This value refers to one of 256 discrete steps in the magnitude range for the file

C.3   Variable Stars

Since version 0.12.2 Stellarium supports the subset of variable stars from GCVS[1] which have HIP identificators (i.e., stars from the Hipparcos catalog). Stellarium's Catalog of Variable Stars is based on the **General Catalogue of Variable Stars** [52].

C.3.1   Variable Star Catalog File Format
General Description

Stellarium's Variable Stars catalog data is kept in the `stars/default` sub-directory of the Installation Directory and/or User Directory (see section 5.1).

File Format

The **gcvs_hip_part.dat** file contains data about 6916 variable stars which have HIP identificators and stored in the plain text format with tab delimiter.

Name	Type	Description
HIP	int	This is the HIP identificator for the star. Used for reference to the main star catalogue.
GCVS	string	This is the GCVS designation for the variable star.
Type	string	Type of variability.
Max	float	Magnitude at maximum brightness.
MFlag	int	Magnitude flag code (0=No flag; 1="(" if max. magnitude is an amplitude; 2="<" if max. magnitude is a bright limit; 3=">" is max. magnitude is a faint limit).
Min I	float	First minimum magnitude or amplitude.
Min II	float	Second minimum magnitude.
V	char	The photometric system for magnitudes.
Epoch	float	Epoch for maximum or minimum light (in Julian days).
Period	float	Period for the variable star (in days).
M-m	float	Rising time or duration of the eclipse (in percent).
Spectrum	string	Spectral class of the variable star.

C.3.2   GCVS Variability Types

An improved system of variability classification is used in the fourth edition of the GCVS, based on recent developments in classification principles and taking into account the suggestions of a number of specialists. Variability types are grouped according to the major astrophysical reasons for variability.

All of these classes include objects of a dissimilar nature that belong to different types of light variability. On the other hand, an object may be variable because of almost all of the possible reasons or because of any combination of them. If a variable belongs to several types of variability, the types are joined in the data field by a "+" sign, e.g., E+UG, UV+BY.

---

[1]http://www.sai.msu.su/gcvs/gcvs/

Despite considerable success in understanding stellar variability processes, the classification adopted in the Catalogue is far from perfect. This is especially the case for explosive, symbiotic and novalike variables; X-ray sources; and peculiar objects.

You can read about latest different types of variability here.

Eruptive Variable Stars

Eruptive variables are stars varying in brightness because of violent processes and flares occurring in their chromospheres and coronae. The light changes are usually accompanied by shell events or mass outflow in the form of stellar winds of variable intensity and/or by interaction with the surrounding interstellar medium. This class includes the following types:

Type	Description
FU	Orion variables of the FU Orionis type. Characterized by gradual increases in brightness by about 6 mag in several months, followed by either almost complete constancy at maximum that is sustained for long periods of time or slow decline by 1-2 mag. Spectral types at maximum are in the range Ae(alpha) - Gpe(alpha). After an outburst, a gradual development of an emission spectrum is observed and the spectral type becomes later. These variables probably mark one of the evolutionary stages of T Tauri-type Orion variables (INT), as evidenced by an outburst of one member, V1057 Cyg, but its decline (2.5 mag in 11 years) commenced immediately after maximum brightness was attained. All presently known FU Ori variables are coupled with reflecting cometary nebulae.
GCAS	Eruptive irregular variables of the Gamma Cas type. These are rapidly rotating B III-IVe stars with mass outflow from their equatorial zones. The formation of equatorial rings or disks is often accompanied by temporary fading. Light amplitudes may reach 1.5 mag in V.
I	Poorly studied irregular variables with unknown features of light variations and spectral types. This is a very inhomogeneous group of objects.
IA	Poorly studied irregular variables of early (O-A) spectral type.
IB	Poorly studied irregular variables of intermediate (F-G) to late (K-M) spectral type.
IN	Orion variables. Irregular, eruptive variables connected with bright or dark diffuse nebulae or observed in the regions of these nebulae. Some of them may show cyclic light variations caused by axial rotation. In the Spectrum-Luminosity diagram, they are found in the area of the main sequence and subgiants. They are probably young objects that, during the course of further evolution, will become light-constant stars on the zero-age main sequence (ZAMS). The range of brightness variations may reach several magnitudes. In the case of rapid light variations having been observed (up to 1 mag in 1-10 days), the letter "S" is added to the symbol for the type (INS). This type may be divided into the following subtypes:
INA	Orion variables of early spectral types (B-A or Ae). They are often characterized by occasional abrupt Algol-like fadings (T Ori);
INB	Orion variables of intermediate and late spectral types, F-M or Fe-Me (BH Cep, AH Ori). F-type stars may show Algol-like fadings similar to those of many INA stars; K-M stars may produce flares along with irregular light variations;

INT,IT	Orion variables of the T Tauri type. Stars are assigned to this type on the basis of the following (purely spectroscopic) criteria: spectral types are in the range Fe-Me. The spectra of most typical stars resemble the spectrum of the solar chromosphere. The feature specific to the type is the presence of the flourescent emission lines Fe II 4046, 4132 A (anomalously intense in the spectra of these stars), emission lines [Si II] and [O I], as well as the absorption line Li I 6707 A. These variables are usually observed only in diffuse nebulae. If it is not apparent that the star is associated with a nebula, the letter "N" in the symbol for the type may be omitted, e.g., IT (RW AUR);
IN(YY)	Some Orion variables (YY Ori) show the presence of absorption components on the redward sides of emission lines, indicating the infall of matter toward the stars' surfaces. In such cases, the symbol for the type may be accompanied by the symbol "YY".
IS	Rapid irregular variables having no apparent connection with diffuse nebulae and showing light changes of about 0.5 - 1.0 mag within several hours or days. There is no strict boundary between rapid irregular and Orion variables. If a rapid irregular star is observed in the region of a diffuse nebula, it is considered an Orion variable and designated by the symbol INS. To attribute a variable to the IS type, it is necessary to take much care to be certain that its light changes are really not periodic. Quite a number of the stars assigned to this type in the third edition of the GCVS turned out to be eclipsing binary systems, RR Lyrae variables, and even extragalactic BL Lac objects.
ISA	Rapid irregular variables of the early spectral types, B-A or Ae;
ISB	Rapid irregular variables of the intermediate and late spectral types, F-M and Fe-Me.
RCB	Variables of the R Coronae Borealis type. These are hydrogen-poor, carbon- and helium-rich, high-luminosity stars belonging to the spectral types Bpe-R, which are simultaneously eruptive and pulsating variables. They show slow nonperiodic fadings by 1-9 mag in V lasting from a month or more to several hundred days. These changes are superposed on cyclic pulsations with amplitudes up to several tenths of a magnitude and periods in the range 30-100 days.
RS	Eruptive variables of the RS Canum Venaticorum type. This type is ascribed to close binary systems with spectra showing Ca II H and K in emission, their components having enhanced chromospheric activity that causes quasi-periodic light variability. The period of variation is close to the orbital one, and the variability amplitude is usually as great as 0.2 mag in V (UX Ari). They are X-ray sources and rotating variables. RS CVn itself is also an eclipsing system (see below).
SDOR	Variables of the S Doradus type. These are eruptive, high-luminosity Bpec-Fpec stars showing irregular (sometimes cyclic) light changes with amplitudes in the range 1-7 mag in V. They belong to the brightest blue stars of their parent galaxies. As a rule, these stars are connected with diffuse nebulae and surrounded by expanding envelopes (P Cyg, Eta Car).
UV	Eruptive variables of the UV Ceti type, these are K Ve-M Ve stars sometimes displaying flare activity with amplitudes from several tenths of a magnitude up to 6 mag in V. The amplitude is considerably greater in the ultraviolet spectral region. Maximum light is attained in several seconds or dozens of seconds after the beginning of a flare; the star returns to its normal brightness in several minutes or dozens of minutes.

UVN	Flaring Orion variables of spectral types Ke-Me. These are phenomenologically almost identical to UV Cet variables observed in the solar neighborhood. In addition to being related to nebulae, they are normally characterized by being of earlier spectral type and greater luminosity, with slower development of flares (V389 Ori). They are possibly a specific subgroup of INB variables with irregular variations superimposed by flares.
WR	Eruptive Wolf-Rayet variables. Stars with broad emission features of He I and He II as well as C II-C IV, O II-O IV, and N III-N V. They display irregular light changes with amplitudes up to 0.1 mag in V, which are probably caused by physical processes, in particular, by nonstable mass outflow from their atmospheres.

Pulsating Variable Stars

Pulsating variables are stars showing periodic expansion and contraction of their surface layers. The pulsations may be radial or nonradial. A radially pulsating star remains spherical in shape, while in the case of nonradial pulsations the star's shape periodically deviates from a sphere, and even neighboring zones of its surface may have opposite pulsation phases.

Depending on the period value, on the mass and evolutionary status of the star, and on the scale of pulsational phenomena, the following types of pulsating variables may be distinguished:

Type	*Description*
ACYG	Variables of the Alpha Cygni type, which are nonradially pulsating supergiants of Bep-AepIa spectral types. The light changes with amplitudes of the order of 0.1 mag often seem irregular, being caused by the superposition of many oscillations with close periods. Cycles from several days to several weeks are observed.
BCEP	Variables of the Beta Cephei type (Beta Cep, Beta CMa), which are pulsating O8-B6 I-V stars with periods of light and radial-velocity variations in the range of 0.1 - 0.6 days and light amplitudes from 0.01 to 0.3 mag in V. The light curves are similar in shape to average radial-velocity curves but lag in phase by a quarter of the period, so that maximum brightness corresponds to maximum contraction, i.e., to minimum stellar radius. The majority of these stars probably show radial pulsations, but some (V469 Per) display nonradial pulsations; multiperiodicity is characteristic of many of these stars.
BCEPS	A short-period group of Beta Cep variables. The spectral types are B2-B3 IV-V; periods and light amplitudes are in the ranges 0.02 - 0.04 days and 0.015 - 0.025 days, respectively, i.e., an order of magnitude smaller than the normally observed ones.
CEP	Cepheids. Radially pulsating, high luminosity (classes Ib-II) variables with periods in the range of 1-135 days and amplitudes from several hundredths to 2 mag in V (in the B band, the amplitudes are greater). Spectral type at maximum light is F; at minimum, the types are G-K. The longer the period of light variation, the later is the spectral type. The maximum of the surface-layer expansion velocity almost coinciding with maximum light.
CEP(B)	Cepheids (TU Cas, V 367 Sct) displaying the presence of two or more simultaneously operating pulsation modes (usually the fundamental tone with the period P0 and the first overtone P1). The periods P0 are in the range from 2 to 7 days, with the ratio P1/P0 approx. 0.71.

CW	Variables of the W Virginis type. These are pulsating variables of the galactic spherical component (old disk) population with periods of approximately 0.8 to 35 days and amplitudes from 0.3 to 1.2 mag in V. They obey a period-luminosity relation different from that for Delta Cep variables (see DCEP). For an equal period value, the W Vir variables are fainter than the Delta Cep stars by 0.7 - 2 mag. The light curves of W Vir variables for some period intervals differ from those of Delta Cep variables for corresponding periods either by amplitudes or by the presence of humps on their descending branches, sometimes turning into broad flat maxima. W Vir variables are present in globular clusters and at high galactic latitudes. They may be separated into the following subtypes:
CWA	W Vir variables with periods longer than 8 days (W Vir);
CWB	W Vir variables with periods shorter than 8 days (BL Her).
DCEP	These are the classical cepheids, or Delta Cep-type variables. Comparatively young objects that have left the main sequence and evolved into the instability strip of the Hertzsprung-Russell (H-R) diagram, they obey the well-known Cepheid period-luminosity relation and belong to the young disk population. DCEP stars are present in open clusters. They display a certain relation between the shapes of their light curves and their periods.
DCEPS	These are Delta Cep variables having light amplitudes <0.5 mag in V (<0.7 mag in B) and almost symmetrical light curves (M-m approx. 0.4 - 0.5 periods); as a rule, their periods do not exceed 7 days. They are probably first-overtone pulsators and/or are in the first transition across the instability strip after leaving the main sequence (SU Cas). Traditionally, both Delta Cep and W Vir stars are quite often called Cepheids because it is often impossible to discriminate between them on the basis of the light curves for periods in the range 3 - 10 days. However, these are distinct groups of entirely different objects in different evolutionary stages. One of the significant spectral differences between W Vir stars and Cepheids is the presence, during a certain phase interval, of hydrogen-line emission in the former and of Ca II H and K emission in the latter.
DSCT	Variables of the Delta Scuti type. These are pulsating variables of spectral types A0-F5 III-V displaying light amplitudes from 0.003 to 0.9 mag in V (usually several hundredths of a magnitude) and periods from 0.01 to 0.2 days. The shapes of the light curves, periods, and amplitudes usually vary greatly. Radial as well as nonradial pulsations are observed. The variability of some members of this type appears sporadically and sometimes completely ceases, this being a consequence of strong amplitude modulation with the lower value of the amplitude not exceeding 0.001 mag in some cases. The maximum of the surface layer expansion does not lag behind the maximum light for more than 0.1 periods. DSCT stars are representatives of the galactic disk (flat component) and are phenomenologically close to the SX Phe variables.
DSCTC	Low amplitude group of Delta Sct variables (light amplitude <0.1 mag in V). The majority of this type's representatives are stars of luminosity class V; objects of this subtype generally are representative of the Delta Sct variables in open clusters.

L	Slow irregular variables. The light variations of these stars show no evidence of periodicity, or any periodicity present is very poorly defined and appears only occasionally. Like for the type I, stars are often attributed to this type because of being insufficiently studied. Many type L variables are really semiregulars or belong to other types.
LB	Slow irregular variables of late spectral types (K, M, C, S); as a rule, they are giants (CO Cyg). This type is also ascribed, in the GCVS, to slow red irregular variables in the case of unknown spectral types and luminosities.
LC	Irregular variable supergiants of late spectral types having amplitudes of about 1 mag in V (TZ Cas).
M	Mira (Omicron) Ceti-type variables. These are long-period variable giants with characteristic late-type emission spectra (Me, Ce, Se) and light amplitudes from 2.5 to 11 mag in V. Their periodicity is well pronounced, and the periods lie in the range between 80 and 1000 days. Infrared amplitudes are usually less than in the visible and may be <2.5 mag. For example, in the K band they usually do not exceed 0.9 mag. If the amplitudes exceed 1 - 1.5 mag , but it is not certain that the true light amplitude exceeds 2.5 mag, the symbol "M" is followed by a colon, or the star is attributed to the semiregular class with a colon following the symbol for that type (SR).
PVTEL	Variables of the PV Telescopii type. These are helium supergiant Bp stars with weak hydrogen lines and enhanced lines of He and C. They pulsate with periods of approximately 0.1 to 1 days, or vary in brightness with an amplitude of 0.1 mag in V during a time interval of about a year.
RR	Variables of the RR Lyrae type, which are radially-pulsating giant A-F stars having amplitudes from 0.2 to 2 mag in V. Cases of variable light-curve shapes as well as variable periods are known. If these changes are periodic, they are called the "Blazhko effect." Traditionally, RR Lyrae stars are sometimes called short-period Cepheids or cluster-type variables. The majority of these stars belong to the spherical component of the Galaxy; they are present, sometimes in large numbers, in some globular clusters, where they are known as pulsating horizontal-branch stars. Like Cepheids, maximum expansion velocities of surface layers for these stars practically coincide with maximum light.
RR(B)	RR Lyrae variables showing two simultaneously operating pulsation modes, the fundamental tone with the period P0 and the first overtone, P1 (AQ Leo). The ratio P1/P0 is approximately 0.745;
RRAB	RR Lyrae variables with asymmetric light curves (steep ascending branches), periods from 0.3 to 1.2 days, and amplitudes from 0.5 to 2 mag in V;
RRC	RR Lyrae variables with nearly symmetric, sometimes sinusoidal, light curves, periods from 0.2 to 0.5 days, and amplitudes not greater than 0.8 mag in V (SX UMa).

RV	Variables of the RV Tauri type. These are radially pulsating supergiants having spectral types F-G at maximum light and K-M at minimum. The light curves are characterized by the presence of double waves with alternating primary and secondary minima that can vary in depth so that primary minima may become secondary and vice versa. The complete light amplitude may reach 3-4 mag in V. Periods between two adjacent primary minima (usually called formal periods) lie in the range 30-150 days (these are the periods appearing in the Catalogue). Two subtypes, RVA and RVB, are recognized:
RVA	RV Tauri variables that do not vary in mean magnitude (AC Her);
RVB	RV Tauri variables that periodically (with periods from 600 to 1500 days and amplitudes up to 2 mag in V) vary in mean magnitude (DF Cyg, RV Tau).
SR	Semiregular variables, which are giants or supergiants of intermediate and late spectral types showing noticeable periodicity in their light changes, accompanied or sometimes interrupted by various irregularities. Periods lie in the range from 20 to >2000 days, while the shapes of the light curves are rather different and variable, and the amplitudes may be from several hundredths to several magnitudes (usually 1-2 mag in V).
SRA	Semiregular late-type (M, C, S or Me, Ce, Se) giants displaying persistent periodicity and usually small (<2.5 mag in V) light amplitudes (Z Aqr). Amplitudes and light-curve shapes generally vary and periods are in the range of 35-1200 days. Many of these stars differ from Miras only by showing smaller light amplitudes;
SRB	Semiregular late-type (M, C, S or Me, Ce, Se) giants with poorly defined periodicity (mean cycles in the range of 20 to 2300 days) or with alternating intervals of periodic and slow irregular changes, and even with light constancy intervals (RR CrB, AF Cyg). Every star of this type may usually be assigned a certain mean period (cycle), which is the value given in the Catalogue. In a number of cases, the simultaneous presence of two or more periods of light variation is observed;
SRC	Semiregular late-type (M, C, S or Me, Ce, Se) supergiants (Mu Cep) with amplitudes of about 1 mag and periods of light variation from 30 days to several thousand days;
SRD	Semiregular variable giants and supergiants of F, G, or K spectral types, sometimes with emission lines in their spectra. Amplitudes of light variation are in the range from 0.1 to 4 mag, and the range of periods is from 30 to 1100 days (SX Her, SV UMa).
SXPHE	Phenomenologically, these resemble DSCT (Delta Sct) variables and are pulsating subdwarfs of the spherical component, or old disk galactic population, with spectral types in the range A2-F5. They may show several simultaneous periods of oscillation, generally in the range 0.04-0.08 days, with variable-amplitude light changes that may reach 0.7 mag in V. These stars are present in globular clusters.
ZZ	ZZ Ceti variables. These are nonradially pulsating white dwarfs that change their brightnesses with periods from 30 s to 25 min and amplitudes from 0.001 to 0.2 mag in V. They usually show several close period values. Flares of 1 mag are sometimes observed; however, these may be explained by the presence of close UV Ceti companions. These variables are divided into the following subtypes:
ZZA	ZZ Cet-type variables of DA spectral type (ZZ Cet) having only hydrogen absorption lines in their spectra;

ZZB	ZZ Cet-type variables of DB spectral type having only helium absorption lines in their spectra.

Rotating Variable Stars

Variable stars with nonuniform surface brightness and/or ellipsoidal shapes, whose variability is caused by axial rotation with respect to the observer. The nonuniformity of surface brightness distributions may be caused by the presence of spots or by some thermal or chemical inhomogeneity of the atmosphere caused by a magnetic field whose axis is not coincident with the rotation axis. These stars are subdivided into the following types:

Type	*Description*
ACV	Alpha2 Canum Venaticorum variables. These are main-sequence stars with spectral types B8p-A7p and displaying strong magnetic fields. Spectra show abnormally strong lines of Si, Sr, Cr, and rare earths whose intensities vary with rotation. They exhibit magnetic field and brightness changes (periods of 0.5-160 days or more). The amplitudes of the brightness changes are usually withine 0.01-0.1 mag in V.
ACVO	Rapidly oscillating Alpha2 CVn variables. These are nonradially pulsating, rotating magnetic variables of Ap spectral type (DO Eri). Pulsation periods are in the range of 6-12 mmag (0.004-0.01 days), while amplitudes of light variation caused by the pulsation are about 0.01 mag in V. The pulsational variations are superposed on those caused by rotation.
BY	BY Draconis-type variables, which are emission-line dwarfs of dKe-dMe spectral type showing quasiperiodic light changes with periods from a fraction of a day to 120 days and amplitudes from several hundredths to 0.5 mag in V. The light variability is caused by axial rotation of a star with a variable degree of nonuniformity of the surface brightness (spots) and chromospheric activity. Some of these stars also show flares similar to those of UV Cet stars, and in those cases they also belong to the latter type and are simultaneously considered eruptive variables.
ELL	Rotating ellipsoidal variables (b Per, Alpha Vir). These are close binary systems with ellipsoidal components, which change combined brightnesses with periods equal to those of orbital motion because of changes in emitting areas toward an observer, but showing no eclipses. Light amplitudes do not exceed 0.1 mag in V.
FKCOM	FK Comae Berenices-type variables. These are rapidly rotating giants with nonuniform surface brightnesses, which have G-K spectral types with broad H and K Ca II emission and sometimes Halpha. They may also be spectroscopic binary systems. Periods of light variation (up to several days) are equal to rotational periods, and amplitudes are several tenths of a magnitude. It is not excluded that these objects are the product of further evolution of EW (W UMa) close binary systems (see below).
PSR	Optically variable pulsars (CM Tau), which are rapidly rotating neutron stars with strong magnetic fields, radiating in the radio, optical, and X-ray regions. Pulsars emit narrow beams of radiation, and periods of their light changes coincide with rotational periods (from 0.004 to 4 s), while amplitudes of the light pulses reach 0.8 mag.

SXARI	SX Arietis-type variables. These are main-sequence B0p-B9p stars with variable-intensity He I and Si III lines and magnetic fields. They are sometimes called helium variables. Periods of light and magnetic field changes (about 1 day) coincide with rotational periods, while amplitudes are approximately 0.1 mag in V. These stars are high-temperature analogs of the ACV variables.

Cataclysmic (Explosive and Novalike) Variables

These are variable stars showing outbursts caused by thermonuclear burst processes in their surface layers (novae) or deep in their interiors (supernovae). We use the term "novalike" for variables that show novalike outbursts caused by rapid energy release in the surrounding space (UG-type stars - see below) and also for objects not displaying outbursts but resembling explosive variables at minimum light by their spectral (or other) characteristics. The majority of explosive and novalike variables are close binary systems, their components having strong mutual influence on the evolution of each star. It is often observed that the hot dwarf component of the system is surrounded by an accretion disk formed by matter lost by the other, cooler, and more extended component. This category is subdivided into the following types:

Type	*Description*
N	Novae. Close binary systems with orbital periods from 0.05 to 230 days. One of the components of these systems is a hot dwarf star that suddenly, during a time interval from one to several dozen or several hundred days, increases its brightness by 7-19 mag in V, then returns gradually to its former brightness over several months, years, or decades. Small changes at minimum light may be present. Cool components may be giants, subgiants, or dwarfs of K-M type. The spectra of novae near maximum light resemble A-F absorption spectra of luminous stars at first. Then broad emission lines (bands) of hydrogen, helium, and other elements with absorption components indicating the presence of a rapidly expanding envelope appear in the spectrum. As the light decreases, the composite spectrum begins to show forbidden lines characteristic of the spectra of gas nebulae excited by hot stars. At minimum light, the spectra of novae are generally continuous or resemble the spectra of Wolf-Rayet stars. Only spectra of the most massive systems show traces of cool components. Some novae reveal pulsations of hot components with periods of approximately 100 s and amplitudes of about 0.05 mag in V after an outburst. Some novae eventually turn out to be eclipsing systems. According to the features of their light variations, novae are subdivided into fast (NA), slow (NB), very slow (NC), and recurrent (NR) categories.
NA	Fast novae displaying rapid light increases and then, having achieved maximum light, fading by 3 mag in 100 or fewer days (GK Per);
NB	Slow novae that fade after maximum light by 3 mag in >= 150 days (RR Pic). Here the presence of the well-known "dip" in the light curves of novae similar to T Aur and DQ Her is not taken into account: The rate of fading is estimated on the basis of a smooth curve, its parts before and after the "dip" being a direct continuation of one another;

NC	Novae with a very slow development and remaining at maximum light for more than a decade, then fading very slowly. Before an outburst these objects may show long-period light changes with amplitudes of 1-2 mag in V (RR Tel); cool components of these systems are probably giants or supergiants, sometimes semiregular variables, and even Mira variables. Outburst amplitudes may reach 10 mag. High excitation emission spectra resemble those of planetary nebulae, Wolf-Rayet stars, and symbiotic variables. The possibility that these objects are planetary nebulae in the process of formation is not excluded;
NL	Novalike variables, which are insufficiently studied objects resembling novae by the characteristics of their light changes or by spectral features. This type includes, in addition to variables showing novalike outbursts, objects with no bursts ever observed; the spectra of novalike variables resemble those of old novae, and small light changes resemble those typical for old novae at minimum light. However, quite often a detailed investigation makes it possible to reclassify some representatives of this highly inhomogeneous group of objects into other types;
NR	Recurrent novae, which differ from typical novae by the fact that two or more outbursts (instead of a single one) separated by 10-80 years have been observed (T CrB).
SN	Supernovae (B Cas, CM Tau). Stars that increase, as a result of an outburst, their brightnesses by 20 mag and more, then fade slowly. The spectrum during an outburst is characterized by the presence of very broad emission bands, their widths being several times greater than those of the bright bands observed in the spectra of novae. The expansion velocities of SN envelopes are in the thousands of km/s. The structure of a star after outburst alters completely. An expanding emission nebula results and a (not always observable) pulsar remains at the position of the original star. According to the light curve shape and the spectral features, supernovae are subdivided into types I and II.
SNI	Type I supernovae. Absorption lines of Ca II, Si, etc., but no hydrogen lines are present in the spectra. The expanding envelope almost lacks hydrogen. During 20-30 days following maximum light, the brightness decreases by approximately 0.1 mag per day, then the rate of fading slows and reaches a constant value of 0.014/day;
SNII	Type II supernovae. Lines of hydrogen and other elements are apparent in their spectra. The expanding envelope consists mainly of H and He. Light curves show greater diversity than those of type I supernovae. Usually after 40-100 days since maximum light, the rate of fading is 0.1 mag per day.

UG	U Geminorum-type variables, quite often called dwarf novae. They are close binary systems consisting of a dwarf or subgiant K-M star that fills the volume of its inner Roche lobe and a white dwarf surrounded by an accretion disk. Orbital periods are in the range 0.05-0.5 days. Usually only small, in some cases rapid, light fluctuations are observed, but from time to time the brightness of a system increases rapidly by several magnitudes and, after an interval of from several days to a month or more, returns to the original state. Intervals between two consecutive outbursts for a given star may vary greatly, but every star is characterized by a certain mean value of these intervals, i.e., a mean cycle that corresponds to the mean light amplitude. The longer the cycle, the greater the amplitude. These systems are frequently sources of X-ray emission. The spectrum of a system at minimum is continuous, with broad H and He emission lines. At maximum these lines almost disappear or become shallow absorption lines. Some of these systems are eclipsing, possibly indicating that the primary minimum is caused by the eclipse of a hot spot that originates in the accretion disk from the infall of a gaseous stream from the K-M star. According to the characteristics of the light changes, U Gem variables may be subdivided into three types: SS Cyg, SU UMa, and Z Cam.
UGSS	SS Cygni-type variables (SS Cyg, U Gem). They increase in brightness by 2-6 mag in V in 1-2 days and in several subsequent days return to their original brightnesses. The values of the cycle are in the range 10 days to several thousand;
UGSU	SU Ursae Majoris-type variables. These are characterized by the presence of two types of outbursts called "normal" and "supermaxima". Normal, short outbursts are similar to those of UGSS stars, while supermaxima are brighter by 2 mag, are more than five times longer (wider), and occur several times less frequently. During supermaxima the light curves show superposed periodic oscillations (superhumps), their periods being close to the orbital ones and amplitudes being about 0.2-0.3 mag in V. Orbital periods are shorter than 0.1 days; companions are of dM spectral type;
UGZ	Z Camelopardalis-type stars. These also show cyclic outbursts, differing from UGSS variables by the fact that sometimes after an outburst they do not return to the original brightness, but during several cycles retain a magnitude between maximum and minimum. The values of cycles are from 10 to 40 days, while light amplitudes are from 2 to 5 mag in V.
ZAND	Symbiotic variables of the Z Andromedae type. They are close binaries consisting of a hot star, a star of late type, and an extended envelope excited by the hot star's radiation. The combined brightness displays irregular variations with amplitudes up to 4 mag in V. A very inhomogeneous group of objects.

Close Binary Eclipsing Systems

We adopt a triple system of classifying eclipsing binary systems: according to the shape of the combined light curve, as well as to physical and evolutionary characteristics of their components. The classification based on light curves is simple, traditional, and suits the observers; the second and third classification methods take into account positions of the binary-system components in the (MV ,B-V) diagram and the degree of inner Roche lobe filling. Estimates are made by applying the simple criteria proposed by Svechnikov and Istomin (1979). The symbols for the types of eclipsing binary systems that we use are given below.

Classification based on the shape of the light curve

Type	Description
E	Eclipsing binary systems. These are binary systems with orbital planes so close to the observer's line of sight (the inclination i of the orbital plane to the plane orthogonal to the line of sight is close to 90 deg) that the components periodically eclipse each other. Consequently, the observer finds changes of the apparent combined brightness of the system with the period coincident with that of the components' orbital motion.
EA	Algol (Beta Persei)-type eclipsing systems. Binaries with spherical or slightly ellipsoidal components. It is possible to specify, for their light curves, the moments of the beginning and end of the eclipses. Between eclipses the light remains almost constant or varies insignificantly because of reflection effects, slight ellipsoidality of components, or physical variations. Secondary minima may be absent. An extremely wide range of periods is observed, from 0.2 to >= 10000 days. Light amplitudes are also quite different and may reach several magnitudes.
EB	Beta Lyrae-type eclipsing systems. These are eclipsing systems having ellipsoidal components and light curves for which it is impossible to specify the exact times of onset and end of eclipses because of a continuous change of a system's apparent combined brightness between eclipses; secondary minimum is observed in all cases, its depth usually being considerably smaller than that of the primary minimum; periods are mainly longer than 1 day. The components generally belong to early spectral types (B-A). Light amplitudes are usually <2 mag in V.
EW	W Ursae Majoris-type eclipsing variables. These are eclipsers with periods shorter than 1 days, consisting of ellipsoidal components almost in contact and having light curves for which it is impossible to specify the exact times of onset and end of eclipses. The depths of the primary and secondary minima are almost equal or differ insignificantly. Light amplitudes are usually <0.8 mag in V. The components generally belong to spectral types F-G and later.

Classification according to the components' physical characteristics

Type	Description
GS	Systems with one or both giant and supergiant components; one of the components may be a main sequence star.
PN	Systems having, among their components, nuclei of planetary nebulae (UU Sge).
RS	RS Canum Venaticorum-type systems. A significant property of these systems is the presence in their spectra of strong Ca II H and K emission lines of variable intensity, indicating increased chromospheric activity of the solar type. These systems are also characterized by the presence of radio and X-ray emission. Some have light curves that exhibit quasi sine waves outside eclipses, with amplitudes and positions changing slowly with time. The presence of this wave (often called a distortion wave) is explained by differential rotation of the star, its surface being covered with groups of spots; the period of the rotation of a spot group is usually close to the period of orbital motion (period of eclipses) but still differs from it, which is the reason for the slow change (migration) of the phases of the distortion wave minimum and maximum in the mean light curve. The variability of the wave's amplitude (which may be up to 0.2 mag in V) is explained by the existence of a long-period stellar activity cycle similar to the 11-year solar activity cycle, during which the number and total area of spots on the star's surface vary.

WD	Systems with white-dwarf components.
WR	Systems having Wolf-Rayet stars among their components (V444 Cyg).

Classification based on the degree of filling of inner Roche lobes

Type	Description
AR	Detached systems of the AR Lacertae type. Both components are subgiants not filling their inner equipotential surfaces.
D	Detached systems, with components not filling their inner Roche lobes.
DM	Detached main-sequence systems. Both components are main-sequence stars and do not fill their inner Roche lobes.
DS	Detached systems with a subgiant. The subgiant also does not fill its inner critical surface.
DW	Systems similar to W UMa systems in physical properties (KW, see below), but not in contact.
K	Contact systems, both components filling their inner critical surfaces.
KE	Contact systems of early (O-A) spectral type, both components being close in size to their inner critical surfaces.
KW	Contact systems of the W UMa type, with ellipsoidal components of F0-K spectral type. Primary components are main-sequence stars and secondaries lie below and to the left of the main sequence in the (MV,B-V) diagram.
SD	Semidetached systems in which the surface of the less massive component is close to its inner Roche lobe.

The combination of the above three classification systems for eclipsers results in the assignment of multiple classifications for object types. These are separated by a solidus ("/") in the data field. Examples are: E/DM, EA/DS/RS, EB/WR, EW/KW, etc.

Optically Variable Close Binary Sources of Strong, Variable X-ray Radiation (X-ray Sources)

Type	Description
X	Close binary systems that are sources of strong, variable X-ray emission and which do not belong to or are not yet attributed to any of the above types of variable stars. One of the components of the system is a hot compact object (white dwarf, neutron star, or possibly a black hole). X-ray emission originates from the infall of matter onto the compact object or onto an accretion disk surrounding the compact object. In turn, the X-ray emission is incident upon the atmosphere of the cooler companion of the compact object and is reradiated in the form of optical high-temperature radiation (reflection effect), thus making that area of the cooler companion's surface an earlier spectral type. These effects lead to quite a peculiar complex character of optical variability in such systems. These objects may be subdivided into the following types:
XB	X-ray bursters. Close binary systems showing X-ray and optical bursts, their duration being from several seconds to ten minutes, with amplitudes of about 0.1 mag in V (V 801 Ara, V 926 Sco);

XF	Fluctuating X-ray systems showing rapid variations of X-ray (Cygnus X-1 = V1357 Cyg) and optical (V821 Ara) radiation on time scalesof dozens of milliseconds;
XI	X-ray irregulars. Close binary systems consisting of a hot compact object surrounded by an accretion disk and a dA - dM-type dwarf. These display irregular light changes on time scales of minutes and hours, and amplitudes of about 1 mag in V. Superposition of a periodic variation because of orbital motion is possible (V818 Sco);
XJ	X-ray binaries characterized by the presence of relativistic jets evident at X-ray and radio wavelengths, as well as in the optical spectrum in the form of emission components showing periodic displacements with relativistic velocities (V1343 Aql);
XND	X-ray, novalike (transient) systems containing, along with a hot compact object, a dwarf or subgiant of G-M spectral type. These systems occasionally rapidly increase in brightness by 4-9 mag in V, in the visible simultaneously with the X-ray range, with no envelope ejected. The duration of the outburst may be up to several months (V616 Mon);
XNG	X-ray, novalike (transient) systems with an early-type supergiant or giant primary component and a hot compact object as a companion. Following the main component's outburst, the material ejected by it falls onto the compact object and causes, with a significant delay, the appearance of X rays. The amplitudes are about 1-2 mag in V (V725 Tau);
XP	X-ray pulsar systems. The primary component is usually an ellipsoidal early-type supergiant. The reflection effect is very small and light variability is mainly caused by the ellipsoidal primary component's rotation. Periods of light changes are between 1 and 10 days; the period of the pulsar in the system is from 1 s to 100 min. Light amplitudes usually do not exceed several tenths of a magnitude (Vela X-1 = GP Vel);
XPR	X-ray pulsar systems featuring the presence of the reflection effect. They consist of a dB-dF-type primary and an X-ray pulsar, which may also be an optical pulsar. The mean light of the system is brightest when the primary component is irradiated by X rays; it is faintest during a low state of the X-ray source. The total light amplitude may reach 2-3 mag in V (HZ Her);
XPRM,XM	X-ray systems consisting of a late-type dwarf (dK-dM) and a pulsar with a strong magnetic field. Matter accretion on the compact object's magnetic poles is accompanied by the appearance of variable linear and circular polarization; hence, these systems are sometimes known as "polars". The amplitudes of the light changes are usually about 1 mag in V but, provided that the primary component is irradiated by X rays, the mean brightness of a system may increase by 3 mag in V. The total light amplitude may reach 4-5 mag in V (AM Her, AN UMa).

If the beam of X-ray emission originating at the magnetic poles of the rotating hot compact object does not pass through the observer's position and the system is not observed as a pulsar, the letter "P" in the above symbols for X-ray-system types is not used. If an X-ray system is also an eclipsing or an ellipsoidal variable, the X-ray symbol is preceded by "E" or "ELL" joined with the X-ray symbol by a "+" sign (e.g., E+X, ELL+X).

Other Symbols

In addition to the variable-star types described above, certain other symbols that need to be explained will be found in the Type data field:

Type	Description
BLLAC	Extragalactic BL Lacertae-type objects. These are compact quasistellar objects showing almost continuous spectra with weak emission and absorption lines and relatively rapid irregular light changes with amplitudes up to 3 mag in V or more. Sources of strong X-ray radiation and radio waves, their emission displays strong and variable linear polarization in the visible and infrared spectral regions. Some objects of this type, considered erroneously to be variable stars and designated in the GCVS system, will probably sometimes be included in the main table of the Catalogue in the future, too.
CST	Nonvariable stars, formerly suspected to be variable and hastily designated. Further observations have not confirmed their variability.
GAL	Optically variable quasistellar extragalactic objects (active galactic nuclei [AGNs]) considered to be variable stars by mistake.
L:	Unstudied variable stars with slow light changes.
QSO	Optically variable quasistellar extragalactic sources (quasars) that earlier were erroneously considered to be variable stars.
S	Unstudied variable stars with rapid light changes.
*	Unique variable stars outside the range of the classifications described above. These probably represent either short stages of transition from one variability type to another or the earliest and latest evolutionary stages of these types, or they are insufficiently studied members of future new types of variables.
+	If a variable star belongs to several types of light variability simultaneously, the types are joined in the Type field by a "+" sign (e.g., E+UG, UV+BY).
:	Uncertainty flag on Type of Variability

The new variability types

The new variability types have been added in the Name-Lists 67- 77 and in the GCVS vol.V.

Type	Description
ZZO	ZZ Cet type variables of the DO spectral type showing HeII and and CIV absorpion lines in their spectra.
AM	AM Her type variables; close binary systems consisting of a dK-dM type dwarf and of a compact object with strong magnetic field, characterized by variable linear and circular polarization of light. The total range of light variations may reach 4-5 mag V.
R	Close binary systems characterized by the presence of strong reflection (re-radiation) of the light of the hot star illuminating the surface of the cooler companion. Light curves are sinusoidal with the period equal to Porb, maximum brightness coinciding with the passage of the hot star in front of the companion. The eclipse may be absent. The range of light variation is about 0.5-1.0mag V (KV Vel).

BE	It becomes more and more clear that, although the majority of Be stars are photometrically variable, not all of them could be properly called GCAS variables. Quite a number of them show small-scale variations not necessarily related to shell events; in some cases the variations are quasi-periodic. By now we are not able to present an elaborated system of classification for Be variables, but we adopt a decision that in the cases when a Be variable cannot be readily described as a GCAS star we give simply BE for the type of variability.
EP	Stars showing eclipses by their planets. Prototype: V0376 Peg.
SRS	Semiregular pulsating red giants with short period (several days to a month), probably high-overtone pulsators. Prototype: AU Ari.
GDOR	Gamma Doradus stars. Early type F dwarfs showing (multiple) periods from several tenths of a day to slightly in excess of one day. Amplitudes usually do not exceed 0.1 mag. Presumably low degree g-mode non-radial pulsators. Prototype: gamma Dor.
RPHS	Very rapidly pulsating hot (subdwarf B) stars. Typical periods are hundreds of seconds, amplitudes are within several hundredths of a magnitude. Prototype: V361 Hya = EC 14026-2647.
LPB	The comparatively long-period pulsating B stars (periods exceeding (LBV) one day).
BLBOO	The so-called "anomalous Cepheids", i.e. stars with periods characteristic of comparatively long-period RRAB variables, but considerably brighter by luminosity (BL Boo = NGC 5466 V19).

C.4 Double Stars

Since version 0.15.0 Stellarium supports the subset of double stars from WDS[2] which have HIP identificators (i.e., stars from the Hipparcos catalog). Stellarium's Catalog of Double Stars is based on the **Washington Double Star Catalog** [33]. Cross reference data for WDS and HIP designations based on XHIP catalog [3] through astronomical database SIMBAD [68].

C.4.1 Double Star Catalog File Format
General Description

Stellarium's Double Stars catalog data is kept in the `stars/default` sub-directory of the Installation Directory and/or User Directory (see section 5.1).

File Format

The **wds_hip_part.dat** file contains data about 16411 double stars which have HIP identificators and stored in the plain text format with tab delimiter.

Name	*Type*	*Description*
HIP	int	This is the HIP identificator for the star. Used for reference to the main star catalogue.
WDS	string	This is the WDS designation for the double star.
Observation	int	Date of last satisfactory observation (in years).
Position Angle	float	Position Angle at date of last satisfactory observation (in decimal degrees).

---

[2]http://ad.usno.navy.mil/wds/

Separation	float	Separation at date of last satisfactory observation (in arcsec).

C.5 Cross-Identification Data

Since version 0.14.0 Stellarium supports cross-identification between stars from Hipparcos (HIP) [3], Smithsonian Astrophysical Observatory (SAO) Star Catalog J2000.0 (SAO) and Henry Draper Catalogue and Extension (HD) [12] catalogues.

C.5.1 Cross-Identification Catalog File Format
General Description

Stellarium's Cross-Identification catalog data is kept in the `stars/default` sub-directory of the Installation Directory and/or User Directory (see section 5.1).

File Format

The **cross-id.dat** file contains cross reference data for catalogues HIP, SAO and HD, which stored in the plain text format with tab delimiter.

Name	Type	Description
HIP	int	This is the HIP identificator for the star. Used for reference to the main star catalogue.
SAO	int	This is the SAO identificator for the star.
HD	int	This is the HD identificator for the star.

D. Configuration Files

D.1 Program Configuration

First, see 5.3 (The Main Configuration File) for information about the file config.ini, including its default installed location, and command line options that can affect how it is processed.

The file config.ini (or a file which you can load instead with the --config <file> option) is structured into the following parts. In addition, plugins can add a section named like the plugin (Exception: The Text User Interface plugin's section is named [tui] for historical reasons).

D.1.1 [astro]

This section includes settings for the commonly displayed objects.

ID	Type	Default	Description
apparent_magnitude_algorithm	string	Harris	Set algorithm for computation of apparent magnitude of the planets. Possible values: *Planesas*, *Mueller*, *Harris* and *Generic*.
nebula_magnitude_limit	float	8.5	Value of limiting magnitude for the deep-sky objects.
star_magnitude_limit	float	6.5	Value of limiting magnitude for the stars. Sometimes you don't want to display more stars when zooming in.
planet_magnitude_limit	float	6.5	Value of limiting magnitude for the planets. Sometimes you don't want to display more planets when zooming in.
flag_nebula_magnitude_limit	bool	false	Set to *true* to activate limit for showing deep-sky objects.

flag_star_magnitude_limit	bool	false	Set to *true* to activate limit for showing stars
flag_planet_magnitude_limit	bool	false	Set to *true* to activate limit for showing planets.
extinction_mode_below_horizon	string	zero	Set extinction mode for atmosphere below horizon. Possible values: *zero*, *mirror* and *max*.
flag_stars	bool	true	Set to *false* to hide the stars on start-up
flag_star_name	bool	true	Set to *false* to hide the star labels on start-up
flag_planets	bool	true	Set to *false* to hide the planet labels on start-up
flag_planets_hints	bool	true	Set to *false* to hide the planet hints on startup (names and circular highlights)
flag_planets_orbits	bool	false	Set to *true* to show the planet orbits on startup
flag_permanent_orbits	bool	false	Set to *true* to show the orbit of planet, when planet is out of the viewport also.
flag_planets_pointers	bool	false	Set to *true* to show the planet pointer markers on startup
flag_ephemeris_markers	bool	true	Set to *true* to show the calculated ephemeris markers for Solar system bodies.
flag_ephemeris_dates	bool	true	Set to *true* to show the dates of calculated ephemeris for Solar system bodies.
flag_light_travel_time	bool	true	Should be *true* to improve accuracy in the movement of the planets by compensating for the time it takes for light to travel. This has a slight impact on performance, but is essential e.g. for Jupiter's moons.
flag_object_trails	bool	false	Turns on and off drawing of object trails (which show the movement of the planets over time).
flag_isolated_trails	bool	true	Turns on and off drawing of isolated trails.
flag_isolated_orbits	bool	true	Turns on and off drawing of isolated orbits.
flag_nebula	bool	true	Set to *false* to hide the nebulae on start-up.

flag_nebula_name	bool	false	Set to *true* to show the nebula labels on start-up.
flag_nebula_display_no_texture	bool	false	Set to *true* to suppress displaying of nebula textures.
nebula_hints_amount	float	3.0	Sets the amount of hints [0...10].
nebula_labels_amount	float	3.0	Sets the amount of labels [0...10].
flag_milky_way	bool	true	Set to *false* to hide the Milky Way.
milky_way_intensity	float	1.0	Sets the relative brightness with which the milky way is drawn. Typical [1...3].
flag_zodiacal_light	bool	true	Set to *false* to hide the zodiacal light
zodiacal_light_intensity	float	1.0	Sets the relative brightness with which the zodiacal light is drawn.
max_mag_nebula_name	float	8.0	Sets the magnitude of the nebulae whose name is shown.
flag_nebula_hints_proportional	bool	false	Enables/disables proportional markers for deep-sky objects.
flag_surface_brightness_usage	bool	false	Enables/disables usage surface brightness for markers for deep-sky objects.
flag_use_type_filter	bool	false	Enables/disables usage of the type filters for deep-sky objects.
flag_nutation	bool	true	Enables/disables usage nutation.
flag_topocentric_coordinates	bool	true	Enables/disables usage topocentric coordinates.
flag_grs_custom	bool	false	Turns on and off to using custom settings for calculation of position of Great Red Spot on Jupiter.
grs_longitude	int	216	Longitude of Great Red Spot on Jupiter in degrees.
grs_drift	float	15.0	Annual drift of Great Red Spot on Jupiter in degrees.
grs_jd	float	2456901.5	Initial JD for calculation of position of Great Red Spot on Jupiter.
grs_measurements_url	string		URL of recent measurements of GRS' position[1].
meteor_zhr	int	10	ZHR for sporadic meteors.
de430_path	string		Path to JPL DE430 ephemerides.
flag_use_de430	bool	false	Enables/disables usage JPL DE430 ephemerides (if available).
de431_path	string		Path to JPL DE431 ephemerides.

[1]Default value is `http://jupos.privat.t-online.de/rGrs.htm`

| flag_use_de431 | bool | false | Enables/disables usage JPL DE431 ephemerides (if available). |

D.1.2 [color]

This section defines the RGB colors for the various objects, lines, grids, labels etc. Values are given in float from 0 to 1. e.g. *1.0,1.0,1.0* for white, or *1,0,0* for red. Leave no whitespace between the numbers!

ID	Default	Colour of the...
default_color	0.5,0.5,0.7	default colour.
azimuthal_color	0.3,0.2,0.1	azimuthal grid.
equatorial_color	0.2,0.3,0.8	equatorial grid (of date).
equatorial_J2000_color	0.1,0.1,0.5	equatorial grid (J2000).
ecliptical_color	0.6,0.3,0.1	ecliptical grid (of date).
ecliptical_J2000_color	0.4,0.1,0.1	ecliptical grid (J2000).
galactic_color	0.0,0.3,0.2	galactic grid.
galactic_equator_color	0.5,0.3,0.1	galactic equator line.
equator_color	0.3,0.5,1.0	equatorial line.
ecliptic_color	0.9,0.6,0.2	ecliptic line (of date).
ecliptic_J2000_color	0.7,0.2,0.2	ecliptic grid (J2000).
meridian_color	0.2,0.6,0.2	meridian line.
horizon_color	0.2,0.6,0.2	horizon line.
const_lines_color	0.2,0.2,0.6	constellation lines.
const_names_color	0.4,0.6,0.9	constellation names.
const_boundary_color	0.3,0.1,0.1	constellation boundaries.
star_label_color	0.4,0.3,0.5	star labels.
cardinal_color	0.8,0.2,0.1	cardinal points.
planet_names_color	0.5,0.5,0.7	planet names.
planet_orbits_color	0.7,0.2,0.2	orbits.
planet_pointers_color	1.0,0.3,0.3	planet pointers.
object_trails_color	1.0,0.7,0.0	planet trails.
precession_circles_color	0.9,0.6,0.2	precession circles.
prime_vertical_color	0.18,0.54,0.18	prime vertical.
longitude_color	0.2,0.4,0.4	opposition/conjunction longitude.
circumpolar_circles_color	0.5,0.5,0.7	circumpolar circles.
colures_color	0.5,0.5,0.7	colures.

telescope_circle_color	0.6,0.4,0.0	telescope location indicator.
telescope_label_color	0.6,0.4,0.0	telescope label (next to location circle).
script_console_keyword_color	1.0,0.0,1.0	syntax highlight for keywords in the script console.
script_console_module_color	0.0,1.0,1.0	syntax highlight for modules in the script console.
script_console_comment_color	1.0,1.0,0.0	syntax highlight for comments in the script console.
script_console_function_color	0.0,1.0,0.0	syntax highlight for functions in the script console.
script_console_constant_color	1.0,0.5,0.5	syntax highlight for constants in the script console.
daylight_text_color	0.0,0.0,0.0	info text at daylight.
dso_label_color	0.2,0.6,0.7	deep-sky objects labels.
dso_circle_color	1.0,0.7,0.2	deep-sky objects symbols, if not of the types below.
dso_galaxy_color	1.0,0.2,0.2	galaxies symbols.
dso_radio_galaxy_color	0.3,0.3,0.3	radio galaxies symbols.
dso_active_galaxy_color	1.0,0.5,0.2	active galaxies symbols.
dso_interacting_galaxy_color	0.2,0.5,1.0	interacting galaxies symbols.
dso_quasar_color	1.0,0.2,0.2	quasars symbols.
dso_possible_quasar_color	1.0,0.2,0.2	possible quasars symbols.
dso_bl_lac_color	1.0,0.2,0.2	BL Lac objects symbols.
dso_blazar_color	1.0,0.2,0.2	blazars symbols.
dso_nebula_color	0.1,1.0,0.1	nebulae symbols.
dso_planetary_nebula_color	0.1,1.0,0.1	planetary nebulae symbols.
dso_reflection_nebula_color	0.1,1.0,0.1	reflection nebulae symbols.
dso_bipolar_nebula_color	0.1,1.0,0.1	bipolar nebulae symbols.
dso_emission_nebula_color	0.1,1.0,0.1	emission nebulae symbols.
dso_dark_nebula_color	0.3,0.3,0.3	dark nebulae symbols.
dso_hydrogen_region_color	0.1,1.0,0.1	hydrogen regions symbols.
dso_supernova_remnant_color	0.1,1.0,0.1	supernovae remnants symbols.
dso_interstellar_matter_color	0.1,1.0,0.1	interstellar matter symbols.
dso_cluster_with_nebulosity_color	0.1,1.0,0.1	clusters associated with nebulosity symbols.
dso_molecular_cloud_color	0.1,1.0,0.1	molecular clouds symbols.

dso_possible_planetary_nebula_color	0.1,1.0,0.1	possible planetary nebulae symbols.
dso_protoplanetary_nebula_color	0.1,1.0,0.1	protoplanetary nebulae symbols.
dso_cluster_color	1.0,1.0,0.1	star clusters symbols.
dso_open_cluster_color	1.0,1.0,0.1	open star clusters symbols.
dso_globular_cluster_color	1.0,1.0,0.1	globular star clusters symbols.
dso_stellar_association_color	1.0,1.0,0.1	stellar associations symbols.
dso_star_cloud_color	1.0,1.0,0.1	star clouds symbols.
dso_star_color	1.0,0.7,0.2	star symbols.
dso_emission_object_color	1.0,0.7,0.2	emission objects symbols.
dso_young_stellar_object_color	1.0,0.7,0.2	young stellar objects symbols.

D.1.3 [custom_selected_info]

You can fine-tune the bits of information to display for the selected object in this section. Set the entry to *true* to display it.

ID	Type	Description
flag_show_absolutemagnitude	bool	absolute magnitude for objects.
flag_show_altaz	bool	horizontal coordinates for objects.
flag_show_catalognumber	bool	catalog designations for objects.
flag_show_distance	bool	distance to object.
flag_show_extra	bool	extra info for object.
flag_show_hourangle	bool	hour angle for object.
flag_show_magnitude	bool	magnitude for object.
flag_show_name	bool	common name for object.
flag_show_radecj2000	bool	equatorial coordinates (J2000) of object.
flag_show_radecofdate	bool	equatorial coordinates (of date) of object.
flag_show_size	bool	size of object.
flag_show_galcoord	bool	galactic coordinates (System II) of object.
flag_show_eclcoord	bool	ecliptic coordinates (J2000 and of date) of object.
flag_show_type	bool	type of object

D.1.4 [custom_time_correction]

Stellarium allows experiments with ΔT. See 16.3.3 for details.

ID	Type	Description
coefficients	[float,float,float]	Coefficients for custom equation of DeltaT
ndot	float	n-dot value for custom equation of DeltaT

year	int	Year for custom equation of DeltaT

D.1.5 [devel]

This section is for developers only.

ID	Type	Description
convert_dso_catalog	bool	Set to *true* to convert file `catalog.txt` into file `catalog.dat`. Default value: *false*.
convert_dso_decimal_coord	bool	Set to *true* to use decimal values for coordinates in source catalog. Default value: *true*.

D.1.6 [dso_catalog_filters]

In this section you can fine-tune which of the deep-sky catalogs should be selected on startup.

ID	Type	Default	Description
flag_show_ngc	bool	true	New General Catalogue (NGC).
flag_show_ic	bool	true	Index Catalogue (IC).
flag_show_m	bool	true	Messier Catalog (M).
flag_show_c	bool	false	Caldwell Catalogue (C).
flag_show_b	bool	false	Barnard Catalogue (B).
flag_show_sh2	bool	false	Sharpless Catalogue (Sh-II).
flag_show_vdb	bool	false	Van den Bergh Catalogue of reflection nebulae (VdB).
flag_show_rcw	bool	false	The RCW catalogue of Hα-emission regions in the southern Milky Way.
flag_show_lbn	bool	false	Lynds' Catalogue of Bright Nebulae (LBN).
flag_show_ldn	bool	false	Lynds' Catalogue of Dark Nebulae (LDN).
flag_show_cr	bool	false	Collinder Catalogue (Cr).
flag_show_mel	bool	false	Melotte Catalogue of Deep Sky Objects (Mel).
flag_show_pgc	bool	false	HYPERLEDA. I. Catalog of galaxies (PGC).
flag_show_ced	bool	false	Cederblad Catalog of bright diffuse Galactic nebulae (Ced).
flag_show_ugc	bool	false	The Uppsala General Catalogue of Galaxies (UGC).

D.1.7 [dso_type_filters]

In this section you can fine-tune which types of the deep-sky objects should be selected on startup.

ID	Type	Default	Description
flag_show_galaxies	bool	true	display galaxies.
flag_show_active_galaxies	bool	true	display active galaxies.
flag_show_interacting_galaxies	bool	true	display interacting galaxies.

flag_show_clusters	bool	true	display star clusters.
flag_show_bright_nebulae	bool	true	display bright nebulae.
flag_show_dark_nebulae	bool	true	display dark nebulae.
flag_show_planetary_nebulae	bool	true	display planetary nebulae.
flag_show_hydrogen_regions	bool	true	display hydrogen regions.
flag_show_supernova_remnants	bool	true	display supernovae remnants.
flag_show_other	bool	true	display other deep-sky objects.

D.1.8 **[gui]**

This section includes settings for the graphical user interface.

ID	Type	Default	Description
base_font_size	int	13	Sets the font size. Typical value: *15*
base_font_name	string	Verdana (Windows) DejaVu Sans (others)	Selects the name for base font
base_font_file	string		Selects the name for font file, e.g. *DejaVuSans.ttf*
flag_show_fps	bool	true	see at how many frames per second Stellarium is rendering
flag_show_fov	bool	true	see how many degrees your vertical field of view is
flag_mouse_cursor_timeout	bool	true	Set to *false* if you want to have cursor visible at all times.
mouse_cursor_timeout	float	10	Set to *0* if you want to keep the mouse cursor visible at all times. non-0 values mean the cursor will be hidden after that many seconds of inactivity
flag_show_flip_buttons	bool	false	Enables/disables display of the image flipping buttons in the main toolbar (see section 4.3.4)
flag_show_nebulae_background_button	bool	false	Set to *true* if you want to have access to the button for enabling/disabling backgrounds for deep-sky objects

selected_object_info	string	all	Values: *all*, *short*, *none*, and *custom* (since V0.12.0; see 4.3.2).
auto_hide_horizontal_toolbar	bool	true	Set to *true* if you want auto hide the horizontal toolbar.
auto_hide_vertical_toolbar	bool	true	Set to *true* if you want auto hide the vertical toolbar.
flag_use_window_transparency	bool	false	If *false*, show menu bars opaque
flag_show_datetime	bool	true	If *true*, display date and time in the bottom bar
flag_time_jd	bool	false	If *true*, use JD format for time in the bottom bar
flag_show_tz	bool	false	If *true*, show time zone info in the bottom bar
flag_show_location	bool	true	If *true*, display location in the bottom bar
flag_fov_dms	bool	false	If *true*, using DMS format for FOV in the bottom bar
flag_show_decimal_degrees	bool	false	If *true*, use decimal degrees for coordinates
flag_use_azimuth_from_south	bool	false	If *true*, calculate azimuth from south towards west (as in some astronomical literature)
flag_show_gui	bool	false	If *true*, display GUI.
flag_dso_designation_usage	bool	false	If *true*, use designations of deep-sky objects on the sky instead of their common names.
flag_enable_kinetic_scrolling	bool	true	If *true*, use kinetic scrolling in the GUI[2].
flag_show_buttons_background	bool	true	If *true*, use background under bottons on bottom panel.
pointer_animation_speed	float	1.0	Animation speed of pointers.

[init_location]

ID	Type	Description
landscape_name	string	Sets the landscape you see. Built-in options are *garching*, *geneva*, *grossmugl*, *guereins*, *hurricane*, *jupiter*, *mars*, *moon*, *neptune*, *ocean*, *saturn*, *trees*, *uranus*, *zero*.

---

[2]This option is available on Windows tablets only.

| location | string | Name of location on which to start stellarium. |
| last_location | string | Coordinates of last used location in stellarium. |

D.1.10 [landscape]

ID	Type	Description
atmosphere_fade_duration	float	Sets the time (seconds) it takes for the atmosphere to fade when de-selected
flag_landscape	bool	Set to false if you don't want to see the landscape at all
flag_fog	bool	Set to *false* if you don't want to see fog on start-up
flag_atmosphere	bool	Set to *false* if you don't want to see atmosphere on start-up
flag_landscape_sets_location	bool	Set to *true* if you want Stellarium to modify the observer location when a new landscape is selected (changes planet and longitude/latitude/altitude if location data is available in the landscape.ini file)
minimal_brightness	float	Set minimal brightness for landscapes. [0...1] Typical value: *0.01*
atmosphereybin	int	Set atmosphere binning coefficient for axis Y.
flag_minimal_brightness	bool	Set to *true* to use minimal brightness for landscape.
flag_landscape_sets_minimal_brightness	bool	Set to *true* to use value for minimal brightness for landscape from landscape settings.
flag_enable_illumination_layer	bool	Set to *true* to use illumination layer for landscape.
flag_enable_labels	bool	Set to *true* to use landscape labels from gazetteer layer.
atmospheric_extinction_coefficient	float	Set atmospheric extinction coefficient k [mag/airmass]
temperature_C	float	Set atmospheric temperature [Celsius]
pressure_mbar	float	Set atmospheric pressure [mbar]

D.1.11 [localization]

ID	Type	Description
sky_culture	string	Sets the sky culture to use. E.g. *western, polynesian, egyptian, chinese, lakota, navajo, inuit, korean, norse, tupi, maori, aztec, sami.*

sky_locale	string	Sets language used for names of objects in the sky (e.g. planets). The value is a short locale code, e.g. *en, de, en_GB*
app_locale	string	Sets language used for Stellarium's user interface. The value is a short locale code, e.g. *en, de, en_GB*.
time_zone	string	Sets the time zone. Valid values: *system_default*, or some region/location combination, e.g. *Pacific/Marquesas*
time_display_format	string	time display format: can be *system_default, 24h* or *12h*.
date_display_format	string	date display format: can be *system_default, mmddyyyy, ddm-myyyy* or *yyyymmdd* (ISO8601).

D.1.12   **[main]**

ID	Type	Description
invert_screenshots_colors	bool	If *true*, Stellarium will saving the screenshorts with inverted colors.
restore_defaults	bool	If *true*, Stellarium will restore default settings at startup. This is equivalent to calling Stellarium with the `-restore-defaults` option.
screenshot_dir	string	Path for saving screenshots
version	string	Version of Stellarium. This parameter may be used to detect necessary changes in config.ini file, do not edit.
use_separate_output_file	bool	Set to *true* if you want to create a new file for script output for each start of Stellarium
ignore_opengl_warning	bool	Set to *true* if you don't want to see OpenGL warnings for each start of Stellarium.
check_requirements	bool	Set to *false* if you want to disable and permanently ignore checking hardware requirements at startup. Expect problems if hardware is below requirements!

D.1.13   **[navigation]**

This section controls much of the look&feel of Stellarium. Be careful if you change something here.

ID	Type	Description
preset_sky_time	float	Preset sky time used by the dome version. Unit is Julian Day. Typical value: *2451514.250011573*
startup_time_mode	string	Set the start-up time mode, can be *actual* (start with current real world time), or *Preset* (start at time defined by preset_sky_time)
flag_enable_zoom_keys	bool	Set to *false* if you want to disable the zoom
flag_manual_zoom	bool	Set to *false* for normal zoom behaviour as described in this guide. When set to true, the auto zoom feature only moves in a small amount and must be pressed many times

flag_enable_move_keys	bool	Set to *false* if you want to disable the arrow keys
flag_enable_mouse_navigation	bool	Set to *false* if you want to disable the mouse navigation.
init_fov	float	Initial field of view, in degrees, typical value: "60".
min_fov	float	Minimal field of view, in degrees, typical value: "0.001389" (5").
init_view_pos	floats	Initial viewing direction. This is a vector with x,y,z-coordinates. x being N-S (S +ve), y being E-W (E +ve), z being up-down (up +ve). Thus to look South at the horizon use *1,0,0*. To look Northwest and up at 45°, use *-1,-1,1* and so on.
auto_move_duration	float	Duration for the program to move to point at an object when the space bar is pressed. Typical value: *1.4*
mouse_zoom	float	Sets the mouse zoom amount (mouse-wheel)
move_speed	float	Sets the speed of movement
zoom_speed	float	Sets the zoom speed
viewing_mode	string	If set to *horizon*, the viewing mode simulate an alt/azi mount, if set to *equator*, the viewing mode simulates an equatorial mount
flag_manual_zoom	bool	Set to *true* if you want to auto-zoom in incrementally.
auto_zoom_out_resets_direction	bool	Set to *true* if you want to auto-zoom restoring direction.
time_correction_algorithm	string	Algorithm of DeltaT correction.

D.1.14 [**plugins_load_at_startup**]

This section lists which plugins are loaded at startup (those with *true* values). Each plugin can add another section into this file with its own content, which is described in the respective plugin documentation, see 10. You activate loading of plugins in the [F2] settings dialog, tab "Plugins". After selection of which plugins to load, you must restart Stellarium.

ID	Type	Description
AngleMeasure	bool	Angle Measure plugin
ArchaeoLines	bool	ArchaeoLines plugin
CompassMarks	bool	Compass Marks plugin
MeteorShowers	bool	Meteor Showers plugin
Exoplanets	bool	Exoplanets plugin
Observability	bool	Observability Analysis
Oculars	bool	Oculars plugin
Pulsars	bool	Pulsars plugin

Quasars	bool	Quasars plugin
RemoteControl	bool	Remote Control plugin
Satellites	bool	Satellites plugin
SolarSystemEditor	bool	Solar System Editor plugin
Supernovae	bool	Historical Supernovae plugin
TelescopeControl	bool	Telescope Control plugin
TextUserInterface	bool	Text User Interface plugin
TimeZoneConfiguration	bool	Time Zone plugin
Novae	bool	Bright Novae plugin
Scenery3dMgr	bool	Scenery 3D plugin

D.1.15 **[projection]**

This section contains the projection of your choice and several advanced settings important if you run Stellarium on a single screen, multi-projection, dome projection, or other setups.

ID	Type	Description
type	string	Sets projection mode. Values: *ProjectionPerspective, ProjectionEqualArea, ProjectionStereographic, ProjectionFisheye, ProjectionHammer, ProjectionCylinder, ProjectionMercator, ProjectionOrthographic, ProjectionMiller,* or *ProjectionSinusoidal.*
flip_horz	bool	
flip_vert	bool	
viewport	string	How the view-port looks. Values: *none* (regular rectangular screen), *disk* (circular screen, useful for planetarium setup).
viewportMask	string	How the view-port looks. Values: *none.*
viewport_fov_diameter	float	
viewport_x	float	Usually 0.
viewport_y	float	Usually 0.
viewport_width	float	
viewport_height	float	
viewport_center_x	float	Usually half of viewport_width.
viewport_center_y	float	Usually half of viewport_height.
viewport_center_offset_x	float	[-0.5...+0.5] Usually 0.
viewport_center_offset_y	float	[-0.5...+0.5] Use negative values to move the horizon lower.

D.1.16 **[proxy]**

This section has setting for connection to network through proxy server (proxy will be using when host of proxy is filled).

ID	Type	Description
host_name	string	Name of host for proxy. E.g. *proxy.org*
type	string	Type of proxy. E.g. *socks*
port	int	Port of proxy. E.g. *8080*
user	string	Username for proxy. E.g. *michael_knight*
password	string	Password for proxy. E.g. *xxxxx*

D.1.17 **[scripts]**

ID	Type	Default	Description
startup_script	string	startup.ssc	name of script executed on program start

D.1.18 **[search]**

ID	Type	Description
flag_search_online	bool	If *true*, Stellarium will be use SIMBAD for search.
simbad_server_url	string	URL for SIMBAD mirror
flag_start_words	bool	If *false*, Stellarium will be search phrase only from start of words
coordinate_system	string	Specifies the coordinate system. *Possible values:* equatorialJ2000, equatorial, horizontal, galactic. *Default value:* equatorialJ2000.

D.1.19 **[spheric_mirror]**

Stellarium can be used in planetarium domes. You can use a projector with a hemispheric mirror with geometric properties given in this section. Note: These functions are only rarely used, some may not work as expected.

ID	Type	Default	Description
flip_horz	bool	true	Flip the projection horizontally
flip_vert	bool	false	Flip the projection vertically
projector_alpha	float	0	This parameter controls the properties of the spheric mirror projection mode.
projector_gamma	float		This parameter controls the properties of the spheric mirror projection mode.
projector_delta	float	-1e100	This parameter controls the properties of the spheric mirror projection mode.
projector_phi	float	0	This parameter controls the properties of the spheric mirror projection mode.

projector_position_x	float	0	
projector_position_y	float	1	
projector_position_z	float	-0.2	
mirror_position_x	float	0	
mirror_position_y	float	2	
mirror_position_z	float	0	
image_distance_div_height	float	-1e100	
mirror_radius	float	0.25	
dome_radius	float	2.5	
custom_distortion_file	string		
texture_triangle_base_length	float		
zenith_y	float	0.125	deprecated
scaling_factor	float	0.8	deprecated
distorter_max_fov	float	175.0	Set the maximum field of view for the spheric mirror distorter in degrees. Typical value: *180*
viewportCenterWidth	float		projected image width, pixels
viewportCenterHeight	float		projected image height, pixels
viewportCenterX	float		projected image center X, pixels
viewportCenterY	float		projected image center Y, pixels

D.1.20 [stars]

This section controls how stars are rendered.

ID	Type	Description
relative_scale	float	relative size of bright and faint stars. Higher values mean that bright stars are comparitively larger when rendered. Typical value: *1.0*
absolute_scale	float	Changes how large stars are rendered. larger value lead to larger depiction. Typical value: *1.0*
star_twinkle_amount	float	amount of twinkling. Typical value: *0.3*
flag_star_twinkle	bool	*true* to allow twinkling (but only when atmosphere is active!).
mag_converter_max_fov	float	maximum field of view for which the magnitude conversion routine is used. Typical value: *90.0*.
mag_converter_min_fov	float	minimum field of view for which the magnitude conversion routine is used. Typical value: *0.001*.
labels_amount	float	amount of labels. Typical value: *3.0*
init_bortle_scale	int	initial value of light pollution on the Bortle scale. Typical value: *3*.

D.1.21 **[tui]**

The built-in text user interface (TUI) plugin (see chapter 11.6) is most useful for planetariums. You can even configure a system shutdown command. For historical reasons, the section is not called [TextUserInterface] but simply [tui].

ID	Type	Default	Description
tui_font_size	float	15	Font size for TUI.
tui_font_color	floatRGB	0.3,1,0.3	Font color for TUI.
flag_show_gravity_ui	bool	false	Enables or disables gravity mode for UI
flag_show_tui_datetime	bool	false	Set to *true* if you want to see a date and time label suited for dome projections
flag_show_tui_short_obj_info	bool		set to *true* if you want to see object info suited for dome projections
tui_admin_shutdown_command	string		e.g. for Linux: "shutdown –poweroff +2"

D.1.22 **[video]**

ID	Type	Description
fullscreen	bool	If *true*, Stellarium will start up in full-screen mode, else windowed mode
screen_w	int	Display width when in windowed mode. Value in pixels, e.g. *1024*
screen_h	int	Display height when in windowed mode. Value in pixels, e.g. *768*
screen_x	int	Horizontal position of the top-left corner in windowed mode. Value in pixels, e.g. *0*
screen_y	int	Vertical position of the top-left corner in windowed mode. Value in pixels, e.g. *0*
viewport_effect	string	This is used when the spheric mirror display mode is activated. Values include *none* and *sphericMirrorDistorter*.
minimum_fps	int	Sets the minimum number of frames per second to display at (hardware performance permitting)
maximum_fps	int	Sets the maximum number of frames per second to display at. This is useful to reduce power consumption in laptops.

D.1.23 **[viewing]**

This section defines which objects, labels, lines, grids etc. you want to see on startup. Set those to *true*. Most items can be toggled with hotkeys or switched in the GUI.

ID	Type	Description
flag_constellation_drawing	bool	Display constellation line drawing
flag_constellation_name	bool	Display constellation names

flag_constellation_art	bool	Display constellation art
flag_constellation_boundaries	bool	Display constellation boundaries
flag_constellation_isolate_selected	bool	If *true*, constellation lines, boundaries and art will be limited to the constellation of the selected star, if that star is "on" one of the constellation lines.
flag_constellation_pick	bool	Set to *true* if you only want to see the line drawing, art and name of the selected constellation star
flag_isolated_trails	bool	Set to *true* if you only want to see the trail line drawn for the selected planet (asteroid, comet, moon)
flag_isolated_orbits	bool	Set to *true* if you want to see orbits only for selected planet and their moons.
flag_azimutal_grid	bool	Display azimuthal grid
flag_equatorial_grid	bool	Display equatorial grid (of date)
flag_equatorial_J2000_grid	bool	Display equatorial grid (J2000)
flag_ecliptic_grid	bool	Display ecliptic grid (of date)
flag_ecliptic_J2000_grid	bool	Display ecliptic grid (J2000)
flag_galactic_grid	bool	Display galactic grid (System II)
flag_galactic_equator_line	bool	Display galactic equator line
flag_equator_line	bool	Display celestial equator line (of date)
flag_equator_J2000_line	bool	Display celestial equator line (J2000)
flag_ecliptic_line	bool	Display ecliptic line (of date)
flag_ecliptic_J2000_line	bool	Display ecliptic line (J2000)
flag_meridian_line	bool	Display meridian line
flag_prime_vertical_line	bool	Display Prime Vertical line (East-Zenith-West)
flag_colure_lines	bool	Display colure lines (Celestial Pole-♈/♋/♎/♑)
flag_cardinal_points	bool	Display cardinal points
flag_gravity_labels	bool	Set to *true* if you want labels to undergo gravity (top side of text points toward zenith). Useful with dome projection.
flag_moon_scaled	bool	Set to *false* if you want to see the real moon size
moon_scale	float	Sets the moon scale factor, sometimes useful to correlate to our perception of the moon's size. Typical value: *4*
constellation_art_intensity	float	brightness [0...1] of the constellation art images. Typical value: *0.5*
constellation_art_fade_duration	float	time the constellation art takes to fade in or out, in seconds. Typical value: *1.5*
constellation_font_size	int	font size for constellation labels

constellation_line_thickness	float	thickness of lines of the constellations [1...5]. Typical value: *1*
flag_night	bool	Enable night mode (red-only mode) on startup
light_pollution_luminance	float	Sets the level of the light pollution simulation
use_luminance_adaptation	bool	Enable dynamic eye adaptation.

D.2 Solar System Configuration File

The file `ssystem.ini` (default file in `data/` in the program directory, or a privately modified copy in your own `data/` directory (see 5) contains orbital and rotational data from which Stellarium configures the Solar System objects. You can edit the file either manually, or with the Solar System Editor plugin (see Section 11.8).

Each object's data are described in its own section which is typically named after the object name. Some section names (e.g. those using diacriticals, spaces or other problematic characters) appear a bit mangled.

We list here examples for the major planets, larger moons with special coordinate functions, minor moons with generic orbital elements, minor planets and comets with elliptical and parabolic orbit elements.

All elements are stored in alphabetic order in the file, however it makes more sense to present the elements in another structure which better reflects the meaning.

D.2.1 Planet section

Example:

```
[jupiter]
name=Jupiter
type=planet
coord_func=jupiter_special

lighting=true
albedo=0.51
atmosphere=1
color=1., 0.983, 0.934
tex_halo=star16x16.png
tex_map=jupiter.png #texture courtesy of Bj\xf6rn J\xf3nsson
halo=true
oblateness=0.064874
orbit_visualization_period=4331.87
parent=Sun
radius=71492
rot_equator_ascending_node=-22.203
rot_obliquity=2.222461
rot_periode=9.92491
rot_pole_de=64.49
rot_pole_ra=268.05
rot_rotation_offset=-1 # use JupiterGRS patch
```

where

name English name of the planet. May appear translated.

type Mandatory for planets:

parent =Sun. The body which this object is running around.

coord_func The planet positions are all computed with a dedicated function (VSOP or DE43x).

orbit_visualization_period number of (earth) days for how long the orbit should be made visible. Typically Stellarium shows one orbit line. The orbit slowly drifts, however.

atmosphere (0 or 1) flag to indicate whether observer locations should enable atmosphere drawing.

radius Equator radius, km.

oblateness Flattening of the polar diameter. (r_{pole}/r_{eq})

lighting Should be true to have shading direction from the sun. The only objects where this should be false is the sun, while comets are often seen to also have false here, to allow rendering a diffuse body without illumination.

albedo total albedo (reflectivity) of the planet. Used for "generic" magnitude computation, but the major planets usually have dedicated magnitude formulas, so the value is not evaluated unless you are observing from a location outside Earth.

color Used to colorize halo

tex_halo File name of texture map for halo texture in `textures` folder. (Used when zoomed out far enough so that the actual sphere is not visible.)

tex_map File name of texture map in `textures` folder.

halo Should be true to draw a halo (simple light disk) when object too small to draw a sphere.

Elements for the Physical Ephemeris

rot_equator_ascending_node *deprecated*

rot_obliquity *deprecated*

rot_pole_ra constant of axis right ascension in ICRF, degrees

rot_pole_ra1 change per century of axis right ascension in ICRF, degrees

rot_pole_de constant of axis declination in ICRF, degrees

rot_pole_de1 change per century of axis declination in ICRF, degrees

rot_rotation_offset longitude of central meridian at J2000.0. The special value here indicates special treatment for the Great Red Spot.

rot_periode Duration of one sidereal rotation, in earth hours.

rot_equator_ascending_node and rot_obliquity are the generic values to specify the axis orientation with respect to the parent object's equator, or the VSOP87 inertial plane in case of planets. However, for planets the IAU has been using right ascension and declination of the pole in the ICRF for decades now. The elements to use for planets should therefore be the rot_pole_* values.

In many cases, the attitude/rotation formula is given like

$$\alpha_0 = \text{rot\_pole\_ra} + T * \text{rot\_pole\_ra1} \tag{D.1}$$

$$\delta_0 = \text{rot\_pole\_de} + T * \text{rot\_pole\_de1} \tag{D.2}$$

$$W = \text{rot\_rotation\_offset} + d * \text{rot\_periode} \tag{D.3}$$

These cases are covered by the data in `ssystem.ini`. Other cases are dealt with in a different way.

D.2.2 Moon section

Moons are special in that they orbit another planet. Therefore, the rotational elements used to be specified relative to the equatorial plane of the parent planet, and `orbit_SemiMajorAxis` in kilometers. However, current IAU reference material gives axis orientation with right ascension and declination values for the pole in ICRF coordinates, with some of them in motion. So again, if one of the `rot_pole_...` values exist, we assume the current standard. For more complicated motion, again some special functions are applied.

```
[amalthea]
name=Amalthea
type=moon
parent=Jupiter

coord_func=ell_orbit
```

```
orbit_AscendingNode=141.5521520794674
orbit_Eccentricity=0.006175744402949701
orbit_Epoch=2454619.50000
orbit_Inclination=0.3864576103404582
orbit_LongOfPericenter=245.4222355150120000
orbit_MeanLongitude=224.7924893552550000
orbit_Period=0.5016370462116355
orbit_SemiMajorAxis=181994.8658358799
orbit_visualization_period=0.5016370462116355

radius=73
rot_equator_ascending_node=213.7
rot_obliquity=15.5
rot_periode=12.039289109079252
rot_rotation_offset=235.50

albedo=0.06
color=1., 0.627, 0.492
halo=true
lighting=true
tex_halo=star16x16.png
tex_map=amalthea.png
```

where

name English name of planet moon. No number, just the name. May be translated.

type moon

parent English name of planet or parent body.

coord_func Must be ell_orbit

orbit_AscendingNode Ω

orbit_Eccentricity e

orbit_Epoch

orbit_Inclination i [degrees]

orbit_LongOfPericenter

orbit_MeanLongitude

orbit_Period

orbit_SemiMajorAxis a [km]

orbit_visualization_period Should be set to orbit_Period to show orbit as closed line. Actually, this happens when omitted. So this is in fact only useful in case of special functions for positioning where orbit_Period is not given.

radius =73

rot_equator_ascending_node =213.7 TRY TO AVOID!

rot_obliquity =15.5 TRY TO AVOID!

rot_pole_ra constant of axis right ascension in ICRF, degrees

rot_pole_ra1 change per century of axis right ascension in ICRF, degrees

rot_pole_de constant of axis declination in ICRF, degrees

rot_pole_de1 change per century of axis declination in ICRF, degrees

rot_rotation_offset longitude of central meridian at J2000.0.

rot_periode Duration of one sidereal rotation, in earth hours.

albedo (actually currently unused when lighting true)

color for drawing halo.

halo =true to draw a simple diffuse dot when zoomed out.

lighting =true to show sphere illuminated.

tex_halo name of halo texture. Should be `star16x16.png`

tex_map name of spherical texture map. Many moons have been mapped by visiting spacecraft!

D.2.3 Minor Planet section

```
[4vesta]
type=asteroid
minor_planet_number=4
name=Vesta

coord_func=comet_orbit
parent=Sun
orbit_ArgOfPericenter=151.19843
orbit_AscendingNode=103.85141
orbit_Eccentricity=0.0887401
orbit_Epoch=2457000.5
orbit_Inclination=7.14043
orbit_MeanAnomaly=20.86389
orbit_MeanMotion=0.27154465
orbit_SemiMajorAxis=2.3617933
orbit_visualization_period=1325.46

color=1., 1., 1.
halo=true
lighting=true
oblateness=0.0

albedo=0.423
radius=280
absolute_magnitude=3.2
slope_parameter=0.32
tex_halo=star16x16.png
tex_map=vesta.png
```

type can be asteroid, dwarf planet, cubewano, plutino, scattered disc object, Oort cloud object. With the exception of Pluto, all positions for minor bodies are computed with the orbiting elements given in this way.

Note that although all minor bodies orbit the sun in an elliptical orbit, the `coord_func=comet_orbit`. However, it is more common for minor planets to specify epoch, mean anomaly and mean motion.

Visual magnitude is modelled from absolute magnitude H and slope parameter G.

Elements for rotational axis may be given just like for planets when they are known. It is recommended to use the modern specification (elements `rot_pole_...`).

D.2.4 Comet section

Comets are tiny, and their outgassing and close approaches to the major planets cause fast changes in their orbital elements, so that each apparition should be specified with a dedicated section in `ssystem.ini`.

The coord_func is specified as `comet_orbit`. Note the specification of a time (JDE) at pericenter which is typical for comets.

Comet brightness is evaluated from

$$\text{mag} = \texttt{absolute\_magnitude} + 5 \cdot \log \text{distance} + 2.5 \cdot \texttt{slope\_parameter} \cdot \log(\text{CometSunDistance}) \tag{D.4}$$

The term `slope_parameter` may be a misnomer in case of comets. From the literature [37] (equation 33.13) we find

$$\text{mag} = g + 5 \log \Delta + \kappa \log r \tag{D.5}$$

from which $\kappa = 2 \cdot \texttt{slope\_parameter}$. In any case, κ is typically $[5 \ldots 15]$ and specific for each comet.

`albedo` is used to set the brightness for rendering the body, if you are close enough. Solar illumination direction in this case is not used.

A large number of elements for historical comets is provided in the file `ssystem_1000comets.ini` in the installation directory. You can copy&paste what you need into your `ssystem.ini`. Unfortunately it is not possible to specify several sets of orbital elements for different epochs which would allow automatic changes.

Periodic Comet

```
[1phalley]
type=comet
name=1P/Halley

coord_func=comet_orbit
parent=Sun
orbit_ArgOfPericenter=111.7154
orbit_AscendingNode=58.8583
orbit_Eccentricity=0.968004
orbit_Inclination=162.2453
orbit_PericenterDistance=0.57136
orbit_TimeAtPericenter=2446463.12979167
orbit_good=78

color=1.0, 1.0, 1.0
dust_brightnessfactor=1.5
dust_lengthfactor=0.4
dust_widthfactor=1.5
lighting=false

albedo=0.1
radius=5
absolute_magnitude=5.5
slope_parameter=3.2
tex_map=nomap.png
```

You may want to e.g. change the name in this entry to name=1P/Halley (1982i). Note a very short configured duration of orbit_good, which means the comet is only displayed 78 days before and after perihel.

Parabolic/Hyperbolic Comet

```
[c2013us10%28catalina%29]
type=comet
name=C/2013 US10 (Catalina)

coord_func=comet_orbit
parent=Sun
orbit_ArgOfPericenter=340.3533
orbit_AscendingNode=186.141
orbit_Eccentricity=1.000372
orbit_Inclination=148.8766
orbit_PericenterDistance=0.822958
orbit_TimeAtPericenter=2457342.20748843
orbit_good=1000

color=1.0, 1.0, 1.0
dust_brightnessfactor=1.5
dust_lengthfactor=0.4
dust_widthfactor=1.5
lighting=false

albedo=0.1
radius=5
absolute_magnitude=4.4
slope_parameter=4
tex_map=nomap.png
```

This has basically the same format. Note eccentricity is larger 1, this means the comet is running on a slightly hyperbolic orbit. Stellarium shows data for this comet for almost 3 years (orbit_good=1000 days) from perihel.

D.2.5 **Solar System Observer**

Stellarium is great for excursions to the surface of any object with known orbital elements. Configuring a viewpoint away from a planet requires a special kind of location. The Solar System Observer has been provided as a neutral view location high above the North pole of the Solar System. It has been configured like this:

```
[solar_system_observer]
name=Solar System Observer
parent=Sun
albedo=0.
color=0., 0., 0.
halo=false
hidden=true
lighting=false
coord_func=ell_orbit
orbit_AscendingNode=0
orbit_Eccentricity=0
orbit_Epoch=2451545.0
```

```
orbit_Inclination=90
orbit_LongOfPericenter=0
orbit_MeanLongitude=90
orbit_Period=70000000000
orbit_SemiMajorAxis=70000000000
radius=1.
rot_obliquity=90
tex_halo=star16x16.png
tex_map=lune.png
```

Note the inclination and mean longitude of 90 degrees.

E. Accuracy

Stellarium originally was developed to present a beautiful simulation of the night sky, mostly to understand what is visible in the sky when you leave your house, i.e., for present times. To save computation time, some concessions were made in astronomical accuracy by using simplified models which seemed acceptable at that time. However, many users started to overstress Stellarium's capabilities to simulate the historical sky of many centuries in the past, and found inconsistencies. Unfortunately, celestial motions are indeed more complicated than a simple clockwork, and the process of retrofitting detailed and accurate models which started around V0.11 is not completed yet. Therefore, when using Stellarium for scientific work like eclipse simulation to illustrate records found in Cuneiform tablets, always also use some other reference to compare. You can of course contact us if you are willing and able to help improving Stellarium's accuracy!

E.1 Planetary Positions

Stellarium uses the VSOP87 [8] theory[1] to calculate the positions of the planets over time.

VSOP87 is an analytical ephemeris modeled to match the numerical integration run DE200 from NASA JPL. Its use is recommended for the years -4000...+8000. You can observe the sun leaving the "ecliptic of date" and running on the "ecliptic J2000" outside this date range. This is obviously a mathematical trick to keep continuity. Still, positions may be somewhat useful outside this range, but don't expect anything reliable 50,000 years in the past!

The optionally usable JPL DE431 delivers planet positions strictly for -13000...+17000 only, and nothing outside. Outside of this range, positions from VSOP87 will be shown again.

As far as Stellarium is concerned, the user should bear in mind the following properties of the VSOP87 method. Accuracy values here are positional as observed from Earth.

---

[1]http://vizier.cfa.harvard.edu/viz-bin/ftp-index?/ftp/cats/VI/81

Object(s)	Method	Notes
Mercury, Venus, Earth-Moon barycenter, Mars	VSOP87	Accuracy is 1 arc-second from 2000 B.C. – 6000 A.D.
Jupiter, Saturn	VSOP87	Accuracy is 1 arc-second from 0 A.D. – 4000 A.D.
Uranus, Neptune	VSOP87	Accuracy is 1 arc-second from 4000 B.C. – 8000 A.D.
Pluto	?	Pluto's position is valid only from 1885 A.D. – 2099 A.D.
Earth's Moon	ELP2000-82B	Unsure about interval of validity or accuracy at time of writing. Possibly valid from 1828 A.D. to 2047 A.D.
Galilean satellites	L2	Valid from 500 A.D. – 3500 A.D.

E.2 Minor Bodies

Positions for the Minor Bodies (Dwarf Planets, Asteroids, Comets) are computed with standard algorithms found in astronomical text books. The generally used method of orbital elements allows to compute the positions of the respective object on an undisturbed Kepler orbit around the sun. However, gravitational, and in the case of comets, non-gravitational (outgassing) disturbances slowly change these orbital elements. Therefore an *epoch* is given for such elements, and computation of positions for times far from this epoch will lead to positional errors. Therefore, when searching for asteroids or comets, always update your orbital elements, and use elements with an epoch as close to your time of observation as possible! Stellarium does not simulate gravitational perturbances and orbital changes of minor bodies passing major planets.

E.3 Precession and Nutation

Since V0.14, Stellarium computes the orientation of earth's axis according to the IAU2006 Precession in a long-time variant developed by Vondrák et al. [67] and IAU2000B Nutation. This also now allows proper depiction of the changes in ecliptic obliquity and display of "instantaneous precession circles" around the ecliptic poles. These circles are indeed varying according to ecliptical obliquity. Nutation is only computed for about 500 years around J2000.0. Nobody could have observed it before 1609, and it is unclear for how long the model is applicable.

E.4 Planet Axes

Orientation for the other planets is still simplified. Future versions should implement modern IAU guidelines.

F. Contributors

When not listed as chapter authors, the following were the main contributors:

Matthew Gates	Primary author of the 1st edition
Paul Robinson	Sky guide; exercise ideas
Andras Mohari	Celestial sphere diagrams; numerous corrections
Rudy Gobits, Dirk Schwarzhans	Mac platform specifics
Barry Gerdes	Windows platform specifics
	Customisation of .fab files
	Documentation Maintainer up to V0.12
Sigma	Japanese translation; many corrections
Richard Powell	Figure 17.3 (colour/magnitude diagram) is a modification of his diagram; he kindly granted permission for it to be distributed under the FDL
John Twin	Many spelling corrections
Georg Zotti	Lead author of the 0.15 edition
Alexander Wolf	New layout, many contributions for the 0.15 edition
The rest of the Stellarium developer team	You know who you are... :-)

Additional material has been incorporated into the guide from sources that are published under the GNU FDL, including material from Wikipedia and the Astronomy book at Wikibooks.

F.1 How you can help

We welcome bug reports, feature requests and feedback through the usual channels (trackers, forums and so on):

Bug tracker `https://bugs.launchpad.net/stellarium`
Questions for Stellarium `https://answers.launchpad.net/stellarium`
Feedback forum `https://sourceforge.net/p/stellarium/discussion/278769/`
Blueprints for Stellarium `https://blueprints.launchpad.net/stellarium`

G. GNU Free Documentation License

Version 1.2, November 2002 Copyright (C) 2000,2001,2002 Free Software Foundation, Inc. 51 Franklin St, Fifth Floor, Boston, MA 02110-1301 USA Everyone is permitted to copy and distribute verbatim copies of this license document, but changing it is not allowed.

G.1 PREAMBLE

The purpose of this License is to make a manual, textbook, or other functional and useful document "free" in the sense of freedom: to assure everyone the effective freedom to copy and redistribute it, with or without modifying it, either commercially or noncommercially. Secondarily, this License preserves for the author and publisher a way to get credit for their work, while not being considered responsible for modifications made by others.

This License is a kind of "copyleft", which means that derivative works of the document must themselves be free in the same sense. It complements the GNU General Public License, which is a copyleft license designed for free software.

We have designed this License in order to use it for manuals for free software, because free software needs free documentation: a free program should come with manuals providing the same freedoms that the software does. But this License is not limited to software manuals; it can be used for any textual work, regardless of subject matter or whether it is published as a printed book. We recommend this License principally for works whose purpose is instruction or reference.

G.2 APPLICABILITY AND DEFINITIONS

This License applies to any manual or other work, in any medium, that contains a notice placed by the copyright holder saying it can be distributed under the terms of this License. Such a notice grants a world-wide, royalty-free license, unlimited in duration, to use that work under the conditions stated herein. The "Document", below, refers to any such manual or work. Any member of the public is a licensee, and is addressed as "you". You accept the license if you copy, modify or distribute the work in a way requiring permission under copyright law.

A "Modified Version" of the Document means any work containing the Document or a portion of it, either copied verbatim, or with modifications and/or translated into another language.

A "Secondary Section" is a named appendix or a front-matter section of the Document that deals exclusively with the relationship of the publishers or authors of the Document to the Document's overall subject (or to related matters) and contains nothing that could fall directly within that overall subject. (Thus, if the Document is in part a textbook of mathematics, a Secondary Section may not explain any mathematics.) The relationship could be a matter of historical connection with the subject or with related matters, or of legal, commercial, philosophical, ethical or political position regarding them.

The "Invariant Sections" are certain Secondary Sections whose titles are designated, as being those of Invariant Sections, in the notice that says that the Document is released under this License. If a section does not fit the above definition of Secondary then it is not allowed to be designated as Invariant. The Document may contain zero Invariant Sections. If the Document does not identify any Invariant Sections then there are none.

The "Cover Texts" are certain short passages of text that are listed, as Front-Cover Texts or Back-Cover Texts, in the notice that says that the Document is released under this License. A Front-Cover Text may be at most 5 words, and a Back-Cover Text may be at most 25 words.

A "Transparent" copy of the Document means a machine-readable copy, represented in a format whose specification is available to the general public, that is suitable for revising the document straightforwardly with generic text editors or (for images composed of pixels) generic paint programs or (for drawings) some widely available drawing editor, and that is suitable for input to text formatters or for automatic translation to a variety of formats suitable for input to text formatters. A copy made in an otherwise Transparent file format whose markup, or absence of markup, has been arranged to thwart or discourage subsequent modification by readers is not Transparent. An image format is not Transparent if used for any substantial amount of text. A copy that is not "Transparent" is called "Opaque".

Examples of suitable formats for Transparent copies include plain ASCII without markup, Texinfo input format, LaTeX input format, SGML or XML using a publicly available DTD, and standard-conforming simple HTML, PostScript or PDF designed for human modification. Examples of transparent image formats include PNG, XCF and JPG. Opaque formats include proprietary formats that can be read and edited only by proprietary word processors, SGML or XML for which the DTD and/or processing tools are not generally available, and the machine-generated HTML, PostScript or PDF produced by some word processors for output purposes only.

The "Title Page" means, for a printed book, the title page itself, plus such following pages as are needed to hold, legibly, the material this License requires to appear in the title page. For works in formats which do not have any title page as such, "Title Page" means the text near the most prominent appearance of the work's title, preceding the beginning of the body of the text.

A section "Entitled XYZ" means a named subunit of the Document whose title either is precisely XYZ or contains XYZ in parentheses following text that translates XYZ in another language. (Here XYZ stands for a specific section name mentioned below, such as "Acknowledgements", "Dedications", "Endorsements", or "History".) To "Preserve the Title" of such a section when you modify the Document means that it remains a section "Entitled XYZ" according to this definition.

The Document may include Warranty Disclaimers next to the notice which states that this License applies to the Document. These Warranty Disclaimers are considered to be included by reference in this License, but only as regards disclaiming warranties: any other implication that these Warranty Disclaimers may have is void and has no effect on the meaning of this License.

G.3 VERBATIM COPYING

You may copy and distribute the Document in any medium, either commercially or noncommercially, provided that this License, the copyright notices, and the license notice saying this License applies to the Document are reproduced in all copies, and that you add no other conditions whatsoever to those of this License. You may not use technical measures to obstruct or control the reading or further copying of the copies you make or distribute. However, you may accept compensation in exchange for copies. If you distribute a large enough number of copies you must also follow the conditions in section 3.

You may also lend copies, under the same conditions stated above, and you may publicly display copies.

G.4 COPYING IN QUANTITY

If you publish printed copies (or copies in media that commonly have printed covers) of the Document, numbering more than 100, and the Document's license notice requires Cover Texts, you must enclose the copies in covers that carry, clearly and legibly, all these Cover Texts: Front-Cover Texts on the front cover, and Back-Cover Texts on the back cover. Both covers must also clearly and legibly identify you as the publisher of these copies. The front cover must present the full title with all words of the title equally prominent and visible. You may add other material on the covers in addition. Copying with changes limited to the covers, as long as they preserve the title of the Document and satisfy these conditions, can be treated as verbatim copying in other respects.

If the required texts for either cover are too voluminous to fit legibly, you should put the first ones listed (as many as fit reasonably) on the actual cover, and continue the rest onto adjacent pages.

If you publish or distribute Opaque copies of the Document numbering more than 100, you must either include a machine-readable Transparent copy along with each Opaque copy, or state in or with each Opaque copy a computer-network location from which the general network-using public has access to download using public-standard network protocols a complete Transparent copy of the Document, free of added material. If you use the latter option, you must take reasonably prudent steps, when you begin distribution of Opaque copies in quantity, to ensure that this Transparent copy will remain thus accessible at the stated location until at least one year after the last time you distribute an Opaque copy (directly or through your agents or retailers) of that edition to the public.

It is requested, but not required, that you contact the authors of the Document well before redistributing any large number of copies, to give them a chance to provide you with an updated version of the Document.

G.5 MODIFICATIONS

You may copy and distribute a Modified Version of the Document under the conditions of sections 2 and 3 above, provided that you release the Modified Version under precisely this License, with the Modified Version filling the role of the Document, thus licensing distribution and modification of the Modified Version to whoever possesses a copy of it. In addition, you must do these things in the Modified Version:

A. Use in the Title Page (and on the covers, if any) a title distinct from that of the Document, and from those of previous versions (which should, if there were any, be listed in the History section of the Document). You may use the same title as a previous version if the original publisher of that version gives permission.

B. List on the Title Page, as authors, one or more persons or entities responsible for authorship of the modifications in the Modified Version, together with at least five of the principal authors of the Document (all of its principal authors, if it has fewer than five), unless they release you from this requirement.

C. State on the Title page the name of the publisher of the Modified Version, as the publisher.

D. Preserve all the copyright notices of the Document.

E. Add an appropriate copyright notice for your modifications adjacent to the other copyright notices.

F. Include, immediately after the copyright notices, a license notice giving the public permission to use the Modified Version under the terms of this License, in the form shown in the Addendum below.

G. Preserve in that license notice the full lists of Invariant Sections and required Cover Texts given in the Document's license notice.

H. Include an unaltered copy of this License.

I. Preserve the section Entitled "History", Preserve its Title, and add to it an item stating at least the title, year, new authors, and publisher of the Modified Version as given on the Title Page. If there is no section Entitled "History" in the Document, create one stating the title, year, authors, and publisher of the Document as given on its Title Page, then add an item describing the Modified Version as stated in the previous sentence.

J. Preserve the network location, if any, given in the Document for public access to a Transparent copy of the Document, and likewise the network locations given in the Document for previous versions it was based on. These may be placed in the "History" section. You may omit a network location for a work that was published at least four years before the Document itself, or if the original publisher of the version it refers to gives permission.

K. For any section Entitled "Acknowledgements" or "Dedications", Preserve the Title of the section, and preserve in the section all the substance and tone of each of the contributor acknowledgements and/or dedications given therein.

L. Preserve all the Invariant Sections of the Document, unaltered in their text and in their titles. Section numbers or the equivalent are not considered part of the section titles.

M. Delete any section Entitled "Endorsements". Such a section may not be included in the Modified Version.

N. Do not retitle any existing section to be Entitled "Endorsements" or to conflict in title with any Invariant Section.

O. Preserve any Warranty Disclaimers.

If the Modified Version includes new front-matter sections or appendices that qualify as Secondary Sections and contain no material copied from the Document, you may at your option designate some or all of these sections as invariant. To do this, add their titles to the list of Invariant Sections in the Modified Version's license notice. These titles must be distinct from any other section titles.

You may add a section Entitled "Endorsements", provided it contains nothing but endorsements of your Modified Version by various parties--for example, statements of peer review or that the text has been approved by an organization as the authoritative definition of a standard.

You may add a passage of up to five words as a Front-Cover Text, and a passage of up to 25 words as a Back-Cover Text, to the end of the list of Cover Texts in the Modified Version. Only one passage of Front-Cover Text and one of Back-Cover Text may be added by (or through arrangements made by) any one entity. If the Document already includes a cover text for the same cover, previously added by you or by arrangement made by the same entity you are acting on behalf of, you may not add another; but you may replace the old one, on explicit permission from the previous publisher that added the old one.

The author(s) and publisher(s) of the Document do not by this License give permission to use their names for publicity for or to assert or imply endorsement of any Modified Version.

G.6 COMBINING DOCUMENTS

You may combine the Document with other documents released under this License, under the terms defined in section 4 above for modified versions, provided that you include in the combination all of the Invariant Sections of all of the original documents, unmodified, and list them all as Invariant Sections of your combined work in its license notice, and that you preserve all their Warranty Disclaimers.

The combined work need only contain one copy of this License, and multiple identical Invariant Sections may be replaced with a single copy. If there are multiple Invariant Sections with the same name but different contents, make the title of each such section unique by adding at the end of it, in parentheses, the name of the original author or publisher of that section if known, or else a unique number. Make the same adjustment to the section titles in the list of Invariant Sections in the license notice of the combined work.

In the combination, you must combine any sections Entitled "History" in the various original documents, forming one section Entitled "History"; likewise combine any sections Entitled "Acknowledgements", and any sections Entitled "Dedications". You must delete all sections Entitled "Endorsements."

G.7 COLLECTIONS OF DOCUMENTS

You may make a collection consisting of the Document and other documents released under this License, and replace the individual copies of this License in the various documents with a single copy that is included in the collection, provided that you follow the rules of this License for verbatim copying of each of the documents in all other respects.

You may extract a single document from such a collection, and distribute it individually under this License, provided you insert a copy of this License into the extracted document, and follow this License in all other respects regarding verbatim copying of that document.

G.8 AGGREGATION WITH INDEPENDENT WORKS

A compilation of the Document or its derivatives with other separate and independent documents or works, in or on a volume of a storage or distribution medium, is called an "aggregate" if the copyright resulting from the compilation is not used to limit the legal rights of the compilation's users beyond what the individual works permit. When the Document is included in an aggregate, this License does not apply to the other works in the aggregate which are not themselves derivative works of the Document.

If the Cover Text requirement of section 3 is applicable to these copies of the Document, then if the Document is less than one half of the entire aggregate, the Document's Cover Texts may be placed on covers that bracket the Document within the aggregate, or the electronic equivalent of covers if the Document is in electronic form. Otherwise they must appear on printed covers that bracket the whole aggregate.

G.9 TRANSLATION

Translation is considered a kind of modification, so you may distribute translations of the Document under the terms of section 4. Replacing Invariant Sections with translations requires special permission from their copyright holders, but you may include translations of some or all Invariant Sections in addition to the original versions of these Invariant Sections. You may include a translation of this License, and all the license notices in the Document, and any Warranty Disclaimers, provided that you also include the original English version of this License and the original versions of those notices and disclaimers. In case of a disagreement between the translation and the original version of this License or a notice or disclaimer, the original version will prevail.

If a section in the Document is Entitled "Acknowledgements", "Dedications", or "History", the requirement (section 4) to Preserve its Title (section 1) will typically require changing the actual title.

G.10 TERMINATION

You may not copy, modify, sublicense, or distribute the Document except as expressly provided for under this License. Any other attempt to copy, modify, sublicense or distribute the Document is void, and will automatically terminate your rights under this License. However, parties who have received copies, or rights, from you under this License will not have their licenses terminated so long as such parties remain in full compliance.

G.11 FUTURE REVISIONS OF THIS LICENSE

The Free Software Foundation may publish new, revised versions of the GNU Free Documentation License from time to time. Such new versions will be similar in spirit to the present version, but may differ in detail to address new problems or concerns. See `http://www.gnu.org/copyleft/`.

Each version of the License is given a distinguishing version number. If the Document specifies that a particular numbered version of this License "or any later version" applies to it, you have the option of following the terms and conditions either of that specified version or of any later version that has been published (not as a draft) by the Free Software Foundation. If the Document does not specify a version number of this License, you may choose any version ever published (not as a draft) by the Free Software Foundation.

Bibliography

[1] S. Kamal Abdali. *The Correct Quibla*. Sept. 1997 (cited on page 137).

[2] Douglas Adams. *The Hitchhiker's Guide to the Galaxy*. Pocket Books, 1981 (cited on page 174).

[3] E. Anderson and C. Francis. "XHIP: An extended hipparcos compilation". In: *Astronomy Letters* 38 (May 2012), pages 331–346. DOI: 10.1134/S1063773712050015. arXiv: 1108.4971 (cited on pages 228, 247, 248).

[4] B. Banjevic. "Ancient eclipses and dating the fall of Babylon". In: *Publications de l'Observatoire Astronomique de Beograd* 80 (May 2006), pages 251–257 (cited on page 179).

[5] E. E. Barnard. *Catalogue of 349 dark objects in the sky*. 1927 (cited on page 82).

[6] K. M. Borkowski. "ELP 2000-85 and the dynamic time-universal time relation". In: *Astronomy and Astrophysics* 205 (Oct. 1988), pages L8–L10 (cited on page 177).

[7] G. Bosch, E. Terlevich, and R. Terlevich. "Gemini/GMOS Search for Massive Binaries in the Ionizing Cluster of 30 Dor". In: *The Astronomical Journal* 137 (Feb. 2009), pages 3437–3441. DOI: 10.1088/0004-6256/137/2/3437. arXiv: 0811.4748 (cited on page 207).

[8] P. Bretagnon and G. Francou. "Planetary theories in rectangular and spherical variables - VSOP 87 solutions". In: *Astronomy and Astrophysics* 202 (Aug. 1988), pages 309–315 (cited on page 275).

[9] Pierre Bretagnon and Jean-Louis Simon. *Planetary Programs and Tables from –4000 to +2800*. Richmond: Willmann-Bell, 1986 (cited on page 177).

[10] D. Brouwer. "A study of the changes in the rate of rotation of the earth". In: *Astronomical Journal* 57 (Sept. 1952), page 125. DOI: 10.1086/106723 (cited on page 177).

[11] Daniel Brown. "Exploring Skyscape in Stellarium". In: *Journal of Skyscape Archaeology* 1.1 (2015), pages 93–111 (cited on page 57).

[12] A. J. Cannon and E. C. Pickering. "VizieR Online Data Catalog: Henry Draper Catalogue and Extension, published in Ann. Harvard Obs. 91-100 (1918-1925)". In: *VizieR Online Data Catalog* 3135 (1993) (cited on page 248).

[13] S. Cederblad. "Studies of bright diffuse galactic nebulae with special regard to their spatial distribution." In: *Meddelanden fran Lunds Astronomiska Observatorium Serie II* 119 (1946), pages 1–166 (cited on page 82).

[14] Michelle Chapront-Touzé and Jean Chapront. *Lunar Tables and Programs from 4000 B.C. to A.D. 8000.* Richmond: Willmann-Bell, 1991 (cited on page 178).

[15] G. M. Clemence. "On the system of astronomical constants". In: *Astronomical Journal* 53 (May 1948), page 169. DOI: 10.1086/106088 (cited on page 177).

[16] P. Collinder. "On Structural Properties of Open Galactic Clusters and their Spatial Distribution. Catalog of Open Galactic Clusters." In: *Annals of the Observatory of Lund* 2 (1931), B1–B46 (cited on page 82).

[17] ESA, editor. *The HIPPARCOS and TYCHO catalogues. Astrometric and photometric star catalogues derived from the ESA HIPPARCOS Space Astrometry Mission.* Volume 1200. ESA Special Publication. 1997 (cited on page 228).

[18] Fred Espenak. *Fifty Year Canon of Solar Eclipses: 1986 – 2035.* Washington: NASA, 1987 (cited on page 177).

[19] Fred Espenak. *Fifty Year Canon of Lunar Eclipses: 1986 – 2035.* Washington: NASA, 1989 (cited on page 177).

[20] Fred Espenak and Jean Meeus. *Five Millennium Canon of Solar Eclipses: -1900 to +3000.* NASA, 2006 (cited on page 178).

[21] *Explanatory Supplement to the Astronomical Ephemeris.* London, 1961. ISBN: 0-11-880578-9 (cited on page 177).

[22] Herman Heine Goldstine. *New and Full Moons 1001 B.C. to A.D. 1651.* Volume 94. American Philosophical Society: Memoirs of the American Philosophical Society. American Philosophical Society, 1973. ISBN: 9780871690944. URL: https://books.google.com/books?id=lgsNAAAAIAAJ (cited on page 177).

[23] G. Henriksson. "A New Test of Einstein's Theory of Relativity by Ancient Solar Eclipses". In: *Cosmology Across Cultures.* Volume 409. Astronomical Society of the Pacific Conference Series. Aug. 2009, page 166 (cited on page 177).

[24] E. Høg et al. "The Tycho-2 catalogue of the 2.5 million brightest stars". In: *Astronomy and Astrophysics* 355 (Mar. 2000), pages L27–L30 (cited on page 228).

[25] M.A. Houlden and F.R. Stephenson. *A Supplement to the Tuckerman Tables.* Volume 170. American Philosophical Society, 1986. ISBN: 9780871691705. URL: https://books.google.ru/books?id=JwsNAAAAIAAJ (cited on page 177).

[26] Sana Islam, Muhammad Sadiq, and Muhammad Shahid Qureshi. "Error Minimization of Polynomial Approximation of Delta T". In: *Journal of Astrophysics and Astronomy* 29 (3-4 2008), pages 363–366. URL: http://www.ias.ac.in/describe/article/joaa/029/03-04/0363-0366 (cited on page 179).

[27] M. Khalid, Mariam Sultana, and Faheem Zaidi. "Delta T: Polynomial Approximation of Time Period 1620–2013". In: *Journal of Astrophysics* 2014 (2014). DOI: 10.1155/2014/480964 (cited on page 179).

[28] S. M. Kwon, S. S. Hong, and J. L. Weinberg. "An observational model of the zodiacal light brightness distribution". In: *New Astronomy* 10 (2004), pages 91–107. DOI: doi:10.1016/j.newast.2004.05.004 (cited on page 194).

[29] B. T. Lynds. "Catalogue of Dark Nebulae." In: *Astrophysical Journal Supplement* 7 (May 1962), page 1. DOI: 10.1086/190072 (cited on page 82).

[30] B. T. Lynds. "Catalogue of Bright Nebulae." In: *Astrophysical Journal Supplement* 12 (Aug. 1965), page 163. DOI: 10.1086/190123 (cited on page 82).

[31] T. Y. Magakian. "Merged catalogue of reflection nebulae". In: *Astronomy and Astrophysics* 399 (Feb. 2003), pages 141–145. DOI: 10.1051/0004-6361:20021743 (cited on page 82).

[32] R. N. Manchester et al. "The Australia Telescope National Facility Pulsar Catalogue". In: *The Astronomical Journal* 129 (Apr. 2005), pages 1993–2006. DOI: 10.1086/428488. eprint: astro-ph/0412641 (cited on page 124).

[33] B. D. Mason et al. "The 2001 US Naval Observatory Double Star CD-ROM. I. The Washington Double Star Catalog". In: *The Astronomical Journal* 122 (Dec. 2001), pages 3466–3471. DOI: 10.1086/323920 (cited on page 247).

[34] J. Meeus and L. Simons. "Polynomial approximations to Delta T, 1620(2000 AD". In: *Journal of the British Astronomical Association* 110 (Dec. 2000) (cited on pages 178, 179).

[35] Jean Meeus. *Astronomical Algorithms*. Richmond: Willmann-Bell, 1991 (cited on page 178).

[36] Jean Meeus. *Mathematical astronomy morsels*. Richmond: Willmann-Bell, 1997. ISBN: 0-943396-51-4 (cited on page 213).

[37] Jean Meeus. *Astronomical Algorithms*. 2nd. Richmond: Willmann-Bell, 1998 (cited on pages 178, 271).

[38] Jean Meeus. *More mathematical astronomy morsels*. Richmond: Willmann-Bell, 2002 (cited on page 213).

[39] Jean Meeus. *Mathematical astronomy morsels III*. Richmond: Willmann-Bell, 2004. ISBN: 0-943396-81-6 (cited on page 213).

[40] P. J. Melotte. "A Catalogue of Star Clusters shown on Franklin-Adams Chart Plates". In: *Memoirs of the Royal Astronomical Society* 60 (1915), page 175 (cited on page 82).

[41] O. Montenbruck and T. Pfleger. *Astronomy on the Personal Computer*. fourth edition. Berlin: Springer, 2000. ISBN: 3-540-67221-4 (cited on page 178).

[42] L. V. Morrison and F. R. Stephenson. "Secular and decade fluctuations in the earth's rotation - 700 BC - AD 1978". In: *Sun and Planetary System*. Edited by W. Fricke and G. Teleki. Volume 96. Astrophysics and Space Science Library. 1982, pages 173–178 (cited on page 177).

[43] L. V. Morrison and F. R. Stephenson. "Historical values of the Earth's clock error ΔT and the calculation of eclipses". In: *Journal for the History of Astronomy* 35 (Aug. 2004), pages 327–336 (cited on pages 178, 179).

[44] L. V. Morrison and F. R. Stephenson. "Addendum: Historical values of the Earth's clock error". In: *Journal for the History of Astronomy* 36 (Aug. 2005), page 339 (cited on page 178).

[45] H. Mucke and J. Meeus. *Canon of Solar Eclipses -2003 to +2526*. Vienna: Astronomisches Buro, 1983 (cited on page 177).

[46] P. M. Muller and F. R. Stephenson. "The accelerations of the earth and moon from early astronomical observations". In: *Growth Rhythms and the History of the Earth's Rotation*. Edited by G. D. Rosenberg and S. K. Runcorn. 1975, pages 459–533 (cited on page 177).

[47] E. Noyola, K. Gebhardt, and M. Bergmann. "Gemini and Hubble Space Telescope Evidence for an Intermediate-Mass Black Hole in ω Centauri". In: *The Astrophysical Journal* 676, 1008-1015 (Apr. 2008), pages 1008–1015. DOI: 10.1086/529002. arXiv: 0801.2782 (cited on page 208).

[48] Anton R. Peters. "Ancient Records Silent on Theory of Relativity. A Refutation of Henriksson's Article." In: *Journal of Cosmology* 9 (2010), pages 2245–2258 (cited on page 177).

[49] Edward M. Reingold and Nachum Dershowitz. *Calendrical Tabulations 1900-2200*. Cambridge: Cambridge University Press, 2002 (cited on page 178).

[50] Edward M. Reingold and Nachum Dershowitz. *Calendrical Calculations*. 3rd. Cambridge: Cambridge University Press, 2007 (cited on page 178).

[51] A. W. Rodgers, C. T. Campbell, and J. B. Whiteoak. "A catalogue of Hα-emission regions in the southern Milky Way". In: *Monthly Notices of the Royal Astronomical Society* 121 (1960), page 103. DOI: 10.1093/mnras/121.1.103 (cited on page 82).

[52] N. N. Samus et al. "General Catalogue of Variable Stars (Samus+ 2007-2015)". In: *VizieR On-line Data Catalog* B/gcvs (2015), page 1 (cited on page 232).

[53] L. D. Schmadel and G. Zech. "Polynomial approximations for the correction delta T E.T.-U.T. in the period 1800-1975". In: *Acta Astronomica* 29 (1979), pages 101–104 (cited on page 177).

[54] L. D. Schmadel and G. Zech. "Empirical Transformations from U.T. to E.T. for the Period 1800-1988". In: *Astronomische Nachrichten* 309 (1988). DOI: 10.1002/asna.2113090316 (cited on page 178).

[55] C. Schoch. "Die säkulare Acceleration des Mondes und der Sonne". In: *Astronomische Abhandlungen. Ergänzungshefte zu den Astronomischen Nachrichten* 8 (1931) (cited on page 177).

[56] S. Sharpless. "A Catalogue of H II Regions." In: *Astrophysical Journal Supplement* 4 (Dec. 1959), page 257. DOI: 10.1086/190049 (cited on page 82).

[57] John P. Snyder. *Map Projections – A Working Manual*. USGS Professional Papers 1395. Washington: U.S. Geological Survey, 1987 (cited on page 39).

[58] H. Spencer Jones. "The rotation of the earth, and the secular accelerations of the sun, moon and planets". In: *Monthly Notices of the Royal Astronomical Society* 99 (May 1939), page 541. DOI: 10.1093/mnras/99.7.541 (cited on page 177).

[59] F. R. Stephenson. "Pre-Telescopic Astronomical Observations". In: *Tidal Friction and the Earth's Rotation*. Edited by P. Brosche and J. Suendermann. 1978, page 5 (cited on page 177).

[60] F. R. Stephenson and L. V. Morrison. "Long-term changes in the rotation of the earth - 700 B.C. to A.D. 1980". In: *Philosophical Transactions of the Royal Society of London Series A* 313 (Nov. 1984), pages 47–70. DOI: 10.1098/rsta.1984.0082 (cited on pages 177, 179).

[61] F. R. Stephenson and L. V. Morrison. "Long-Term Fluctuations in the Earth's Rotation: 700 BC to AD 1990". In: *Philosophical Transactions of the Royal Society of London Series A* 351 (Apr. 1995), pages 165–202. DOI: 10.1098/rsta.1995.0028 (cited on page 178).

[62] F. Richard Stephenson. *Historical Eclipses and Earth's Rotation*. Cambridge Books Online. Cambridge University Press, 1997. ISBN: 9780511525186. DOI: 10.1017/CBO9780511525186 (cited on page 178).

[63] Bryant Tuckerman. *Planetary, Lunar, and Solar Positions, 601 B.C. to A.D. 1, at Five-day and Ten-day Intervals*. Volume 56. American Philosophical Society: Memoirs of the American Philosophical Society. American Philosophical Society, 1962. URL: https://books.google.com/books?id=ZyUWAAAAIAAJ (cited on page 177).

[64] Bryant Tuckerman. *Planetary, Lunar and Solar Positions A.D. 2 to A.D. 1649 at Five Day and Ten Day Intervals*. Volume 59. American Philosophical Society: Memoirs of the American Philosophical Society. American Philosophical Society, 1964. ISBN: 9780871690593. URL: https://books.google.com/books?id=cDQLAAAAIAAJ (cited on page 177).

[65] S. van den Bergh. "A study of reflection nebulae." In: *Astronomical Journal* 71 (Dec. 1966), pages 990–998. DOI: 10.1086/109995 (cited on page 82).

[66] M.-P. Véron-Cetty and P. Véron. "A catalogue of quasars and active nuclei: 13th edition". In: *Astronomy and Astrophysics* 518, A10 (July 2010), A10. DOI: 10.1051/0004-6361/201014188 (cited on page 126).

[67] J. Vondrák, N. Capitaine, and P. Wallace. "New precession expressions, valid for long time intervals". In: *Astronomy and Astrophysics* 534.A22 (2011), pages 1–19. DOI: 10.1051/0004-6361/201117274 (cited on pages 173, 276).

[68] M. Wenger et al. "The SIMBAD astronomical database. The CDS reference database for astronomical objects". In: *Astronomy and Astrophysics Supplement* 143 (Apr. 2000), pages 9–22. DOI: 10.1051/aas:2000332. eprint: astro-ph/0002110 (cited on pages 82, 247).

[69] N. Zacharias et al. "The Naval Observatory Merged Astrometric Dataset (NOMAD)". In: *American Astronomical Society Meeting Abstracts*. Volume 36. Bulletin of the American Astronomical Society. Dec. 2004, page 1418 (cited on page 228).

[70] Georg Zotti. "Visualization Tools and Techniques". In: *Handbook for Archaeoas-tronomy and Ethnoastronomy*. Edited by Clive L.N. Ruggles. Volume 1. New York: Springer Reference, 2015. Chapter 29, pages 445–457 (cited on page 148).

[71] Georg Zotti. "Archaeoastronomical Simulations in a Desktop Planetarium". In: *Proc. CHNT2015*. Edited by Wolfgang Börner. TO APPEAR. 2016 (cited on page 148).

[72] Georg Zotti. "Virtual Archaeoastronomy". In: *Proc. SEAC2015, Rome*. Edited by Vito F. Polcaro et al. to appear in MAA. 2016 (cited on pages 135, 148).

[73] Georg Zotti and Wolfgang Neubauer. "A Virtual Reconstruction Approach for Archaeoastronomical Research". In: *Proc. VSMM2012 (Virtual Systems in the Infor-mation Society)*. Edited by Gabriele Guidi and Alonzo C. Addison. IEEE. Milano: IEEE, Sept. 2012, pages 33–40. ISBN: 978-1-4673-2563-9 (cited on page 148).

[74] Georg Zotti and Wolfgang Neubauer. "Virtual Reconstructions in a Desktop Planetar-ium for Demonstrations in Cultural Astronomy". In: *Progress in Cultural Heritage Preservation (Proc. 4th International Conference EuroMed 2012, Limassol, Cyprus, Oct. 29–Nov. 3, 2012)*. Edited by M. Ioannides et al. Volume 7616. Lecture Notes in Computer Science. Heidelberg: Springer, Nov. 2012, pages 170–180. ISBN: 978-3-642-34233-2 (cited on page 148).

[75] Georg Zotti and Wolfgang Neubauer. "Astronomical and Topographical Orientation of Kreisgrabenanlagen in Lower Austria". In: *SEAC2011 Stars and Stones: Voyages in Archaeoastronomy and Cultural Astronomy*. Edited by Fernando Pimenta et al. BAR International Series 2720. Oxford: Archaeopress, 2015, pages 188–193. ISBN: 978 1 4073 1373 3 (cited on page 57).

[76] Georg Zotti and Günther Wuchterl. "Raising Awareness on Nocturnal Light Pol-lution around Astronomical Cultural Heritage Sites". In: *Proc. SEAC2014, Malta*. Edited by Tore Lomsdalen et al. to appear. 2016 (cited on page 78).

Index

A

absolute magnitude 180
air mass 197
Almagest 173
altitude 170
Altitude/Azimuth 170
Andromeda Galaxy 211
apparent magnitude 180
apparent radiant 128
Atmospheric Extinction 197
Atmospheric Refraction 197
azimuth 170

B

Barnard's Star 183
Bayer 187
binoculars 211
brown dwarfs 189

C

cardinal points 171
Celestial Equator 170
Celestial Poles 170
Celestial Sphere 169
Charles Messier 195
crossbow 179

D

DE430 51
DE431 51
declination 172
Deep-sky objects 39
Douglas Adams 174
Downloading 22
dwarf stars 189

E

ecliptic 173
ecliptical latitude 173
ecliptical longitude 173
Edmund Halley 193
Ejnar Hertzsprung 190
Equatorial Coordinate System 171
equinox 173
Ernst Hammer 40

F

FAQ 23
First Point of Aries 172
Flamsteed 187

G

Galactic Coordinates 174

galactic latitude 174
galactic longitude 174
giants 189

H

Handy Angles 179, 211
Henry Norris Russell 190
Hipparch 173
Hipparchus 180, 187
Horizontal Coordinate System 170

I

Installation 22

J

Johann Bayer 187
John E. Bortle 223
John Flamsteed 187
John Herschel 175
Julian Day 32, 175

L

Large Magellanic Cloud 206
Linux 22, 23
luminosity class 189

M

M31 211
main sequence 189
mathematical horizon 170
meridian convergence 146
meteor shower 128
meteoroids 128
Modified Julian Date 176
Modified Julian Day 32

N

nadir 170
Norman Pogson 180
Northern Ecliptical Pole 173

O

obliquity of the ecliptic 181

Opposition/Conjunction Line 173
OS X 22, 23
Osborn Maitland Miller 40

P

planetary nebulae 174
population index 129
precession 173, 181
precession circles 173
Proper motion 183
Ptolemy 173, 187
pulsars 174

R

radiant 128
Right ascension 172
Right Ascension/Declination 171
Running Stellarium 22

S

sidereal time 175
Small Magellanic Cloud 207
spectral types 189
spectrum 188
super-giants 189
System Requirements 21

T

time dragging 27
Tycho Brahe 173

V

VSOP87 51

W

white dwarfs 189
Windows 22

Z

zenith 170
Zenithal Hourly Rate 129